Peer-Led Team Learning

A Handbook for Team Leaders

The Workshop Project

Sponsored by the National Science Foundation

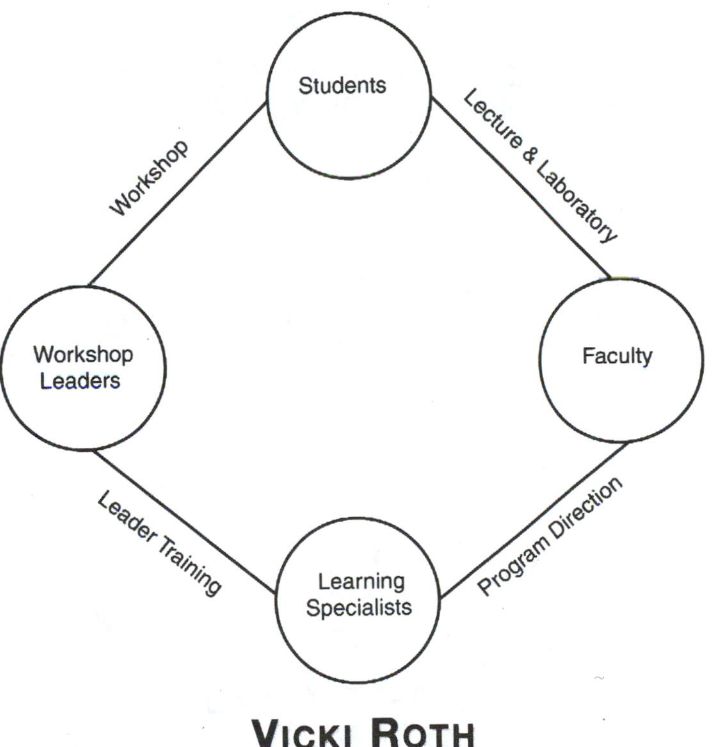

Vicki Roth
Ellen Goldstein
Gretchen Marcus

Prentice Hall Series in Educational Innovation

PRENTICE HALL, Upper Saddle River, NJ 07458

Executive Editor: ***John Challice***
Project Manager: ***Kristen Kaiser***
Editorial Assistant: ***Eliana Ortiz***
Special Projects Manager: ***Barbara A. Murray***
Production Editor: ***Blake Cooper***
Manufacturing Manager: ***Trudy Pisciotti***
Manufacturing Buyer: ***Lisa McDowell***
Supplement Cover Manager: ***Jayne Conte***
Supplement Cover Designer: ***Maureen Eide***

Printed in the United States of America

10 9 8 7 6 5 4 3 2 1

ISBN 0-13-040811-5

Prentice-Hall International (UK) Limited, *London*
Prentice-Hall of Australia Pty. Limited, *Sydney*
Prentice-Hall Canada, Inc., *Toronto*
Prentice-Hall Hispanoamericana, S.A., *Mexico*
Prentice-Hall of India Private Limited, *New Delhi*
Pearson Education Asia Pte. Ltd., Singapore
Prentice-Hall of Japan, Inc., *Tokyo*
Editora Prentice-Hall do Brasil, Ltda., *Rio de Janeiro*

Preface to the Peer-Led Team Learning Series

The Workshop Chemistry Project was an exploration, development, and application of the concept of peer-led team learning in problem-solving Workshops in introductory chemistry courses. A pilot project was first supported by the National Science Foundation, Division of Undergraduate Education, in 1991. In 1995, the Workshop Chemistry Project was selected by NSF/DUE as one of five systemic initiatives to "change the way introductory chemistry is taught." In the period 1991-1998, the project grew from the initial explorations at the City College of New York to a national activity involving more than 50 faculty members at a diverse group of more than 30 colleges and universities. In 1998-1999, approximately 2500 students were guided in Workshop courses by 300 peer leaders per term. In Fall 1999, NSF chose the Workshop Project for a National Dissemination Grant to substantially broaden the chemistry participation and to extend the model to other SMET disciplines, including biology, physics and mathematics.

Peer-Led Team Learning: A Guidebook is the first of a series of five publications that report the work of the Project during the systemic initiative award (1995-1999). The purpose of these five books is to lower the energy barrier to new implementations of the model. The *Guidebook* is a comprehensive account that works back and forth from the conceptual and theoretical foundations of the model to reports of "best-practice" implementation and application. Three other books provide specific materials for use in Workshops in *General Chemistry*; *Organic Chemistry*; and *General, Organic and Biochemistry*. One book in the series, *On Becoming a Peer Leader*, provides materials for leader training.

The collaboration of students, faculty, and learning specialists is a central feature of the Workshop model. The project has been enriched by the talents and energy of many participants. Some of their names are found throughout these books; many others are not identified. In either case, we are most grateful to all those who have advanced the model by their keen insight and enthusiastic commitment.

We also acknowledge, with pleasure, the support of the National Science Foundation, NSF/DUE 9450627 and NSF/DUE 9455920. Our work on the second NSF award was skillfully guided by our National Visiting Committee, Michael Gaines, Chair; Joseph Casanova; Patricia Cuniff; David Evans; Eli Fromm; John Johnson; Bonnie Kaiser; Clark Landis; Kathleen Parson; Arlene Russell; Frank Sutman; Jeffrey Steinfeld; and Ronald Thornton; we value their advice and encouragement. The text of the *Guidebook* was repeatedly processed by Arlene Bristol, with exceptional skill and remarkable patience. Finally, we appreciate the vision and commitment of John Challice and Prentice Hall to make this work readily available to a large audience.

Books are written for you, the readers. We welcome your comments and insights. Please contact us at the indicated e-mail addresses.

The Editors

David K. Gosser gosser@scisun.sci.ccny.cuny.edu
Mark S. Cracolice markc@selway.umt.edu
J. A. Kampmeier kamp@chem.rochester.edu
Vicki Roth vrth@mail.rochester.edu
Victor S. Strozak vstrozak@gc.cuny.edu
Pratibha Varma-Nelson varmanelson@sxu.edu

Introduction

A Word to the Instructor

The Workshop model is a particular way to help students learn; it has its own distinctive structure, characteristics, and style. At the center of the Workshop is the peer leader. Although faculty and learning specialists help to shape the structure and format of the Workshop, it is the peer leader who shapes the interpersonal relationships and intrapersonal attitudes that transform a random group of students into a high-performance team. The peer leader needs to learn how to do this job. Although some elements of the Workshop are common to more familiar learning mechanisms, the unique features of the Workshop guarantee that the peer leaders will have to learn new techniques and maybe even unlearn some old ones.

One way or another, you need to help the peer leaders learn to play their new roles. This is a challenge and an opportunity. The challenge is that we are usually more secure in our disciplines than in pedagogical theory and group practice. That is, simultaneously, the opportunity. This leader training handbook is written for the "peer leader in training." But, it also provides you with a helpful syllabus and appropriate content for leader training. The ideas and the substance are robust and can be adapted to a variety of different formats for peer leader training. At the same time, the handbook provides you a tour through some basic ideas about how students learn. This is a stimulating invitation to reflect on your own pedagogical practices and ideas. Some of the ideas will ring the bell of recognition and provide the basis for the techniques that you have discovered along the way; other ideas will be new and open up pathways for change and innovation.

Jack Kampmeier, Department of Chemistry, University of Rochester

Table of Contents

AN ANTHOLOGY OF READINGS

Reading for Chapter One

Readings for Chapter Two

Readings for Chapter Three

Reading for Chapter Four

Readings for Chapter Five

Readings for Chapter Six

Readings for Chapter Seven

Readings for Chapter Eight

APPENDIX

Education is not the filling of a pail,

but the lighting of a fire.

Yeats

Chapter One: The Workshop Philosophy

If you are reading this chapter, chances are you will be—or you are considering—taking on the role of a workshop leader. If this is true for you, congratulations! Being recruited as a workshop leader is strong testimony to the quality of your previous academic performance and your ability to work with others. We also congratulate you on the opportunity to work with your own group of workshop students. The first time you coach them through a complex problem and see the lights turn on in their eyes, you'll see why we are so excited about this approach to teaching and learning.

Those of us who have been developing the peer-led team learning model would like to thank you in advance for your efforts. The workshop model is a student-centered format; if it weren't for the work of hundreds of previous student leaders, this approach to learning would never have left the ground. And without *your* contributions, this model would have nowhere to go. Your participation, and your feedback, are absolutely crucial, so we've included information about how to contact us at the back of this book. We'd like to hear about your experiences in leading workshops and about improving this handbook for future leaders.

What is a workshop?

If you haven't been in a workshop before, you may be wondering what this model is all about. The idea is really simplicity itself. Small groups of students get together under the guidance of a trained student (that's you) who has been successful in the course in a previous semester. Under the gentle mentoring of the leader, the group works its way through a set of challenging problems. The topics of these problems are introduced and reinforced during regular lectures. One goal is for the student participants to construct their own genuine knowledge of the discipline by working through real issues. Another is for all of the students in the group to feel that they have gained experience and confidence in tough problem solving each week—and that they've achieved this success together.

Sounds easy in principle. But in practice, getting a group to work smoothly for a whole semester requires genuine skill and finesse. If you've been nominated to be a leader, this is an indication that you have much of the natural talent needed to set a group in motion and keep it functioning well; this handbook is designed to help you make the most of your own intrinsic abilities.

Here is the cast of characters often involved in the workshop model: the instructors, who teach the lecture and prepare or modify the workshop materials; learning specialists, who participate with the instructor in leader training; the peer leaders; and the workshop participants themselves.

A very brief history of workshops

The workshop model that you will be learning about in this book has a considerable history. This model, and its predecessors, have their origin in a basic dissatisfaction with certain elements of traditional instructional formats. Remember the old saying about necessity being the mother of invention? That is what has been at work here. For a number of years, instructors, administrators, and especially students themselves, have been less than happy with exclusively lecture-focused courses. The dropout and failure rate in many large lecture courses, and the subsequent shrinkage in the number of students majoring in the natural sciences have been unacceptable (Iadevaia, 1989; Swager, 1995; Seymour, 1997). If we are to do well as a culture, we need as many of us as possible to find success in these tough courses. So we are committed to making these subjects more accessible, without sacrificing any of the complexities of our disciplines.

This isn't to say that lectures themselves are an inherently poor approach; in fact, for some components of the learning cycle, they are nearly ideal. They are a great way to receive a lot of complicated information directly from experts in the field, so lectures are very much part of the workshop model. But, even in the hands of the best lecturer, this format doesn't provide much opportunity for students to discuss and debate those little questions that are often the beginning steps to understanding, or to practice solving problems under someone's guidance, or to teach other people what they've learned. You've probably heard, and maybe experienced yourself, that real learning happens when you help someone else work through complex material. This is hard for students to do in the middle of a lecture, but it is at the very heart of the workshop model.

Although this format has distinctive features of its own, we would like to note our debt to Johnson and Johnson (1991; Johnson, 1994) and Slavin (1982, 1991). Their work on developing cooperative learning formats for the classroom has been of real benefit to us as we have refined our approach. They demonstrated to all levels of academia that students can do well with the tasks involved in running groups (i.e., students don't need someone babysitting their every learning activity) and that group learning is a viable, even necessary, part of learning in many subject areas. The workshop model also owes much to the work of Uri Treisman, who developed a cooperative learning model for the study of calculus (1992). Currently, other faculty and staff are exploring further the path laid down by these earlier researchers in cooperative learning (see, for example, the work of Hanson and Wolfskill, 1997).

Our own approach to this model began in the middle 1990s, when a number of faculty members and college administrators organized a consortium of institutions to work on better learning environments for chemistry. With the support of the National Science Foundation and the administrations of several institutions, the development of the model began. (For more information, check *The PLTL Workshop Project* website listed in the Acknowledgments section.)

Imagine what that first year was like! We were busy completing problems just a few days before each workshop session, and our brave leaders stepped right in to pilot groups within a format they had never experienced themselves. Luckily, we had a lot of good sports involved, most especially the workshop participants themselves. As a result, a great deal of progress was made that first year. From the start, students, leaders, faculty members, and staff saw promise in this model, so it was easy to find the motivation to continue our work.

We have wanted to be sure that this model works for a wide range of students; so we have included all types of institutions in the development of workshops. This format has been tested at research universities, urban commuter campuses, community colleges, technical colleges, state universities, and small liberal arts campuses. Each environment offers its own particular advantages to the installation of the model, and each presents its own obstacles as well. For example, at a community college, it is often a challenge to schedule workshop sessions, since commuter students often have considerable family and work responsibilities elsewhere. On the other hand, the real-life experience community college students bring to the workshops confers an advantage to the group process. At a research university, it is sometimes hard for leaders to find time to devote to their workshop sessions, given the phenomenal number of hours many of them invest in their labs each week, but, in turn, their experience is a true benefit to their workshop participants.

We have also wanted to be sure that this model is useful for a broad range of disciplines. You may be reading this because you will be a leader in a chemistry group, or you may be involved in a biology, physics, or math program. As this is being written, we are investigating the value of this model for social science courses as well, so you could be reading this in preparation for leading groups for courses like economics or linguistics.

A closer look at the model

Now let's spend a few moments outlining the nature of workshops in a little more detail. At the risk of sounding negative, we'll start out by discussing what a workshop *isn't*. A workshop isn't a reprise of the lecture; we aren't asking leaders to stand up and repeat what the instructor has already said about the subject matter. Certainly leaders will want to know and understand what was discussed and demonstrated in class, but the goal of the workshop definitely isn't to re-do.

A workshop isn't a recitation or question-and-answer session, where students bring their problems *du jour* so the leader can fill in the blanks. Of course, questions are asked during the session, but the intention of workshops is to work collaboratively to solve problems, not to ask, say, for few pointers about Problem #2 on page 39 of the text. A workshop isn't a tutorial, either, during which one person serves as the "helper" and the others are the "helpee's." In workshops, the intent is for everyone to help each other with the material—and to avoid a passive or remedial approach to this work. This model has been shown to benefit students of all levels of ability and experience, so we want the whole group to feel actively engaged by these sessions.

And finally, while this model borrows a great deal from the Supplementary Instruction (SI) format (Martin & Burmeister, 1996; Reittinger & Palmer, 1996), workshops are distinctly different. We especially admire the use of student leaders in SI, but our model more closely marries the sessions themselves to the curriculum and to the faculty of the course.

So now that we've reviewed what a workshop isn't, we should discuss what it *is*. There is no such thing as an ideal workshop, but here are some general ideas of what happens in a successful program.

How many, how often

The class is divided into small sets of students, usually six to eight per group. Your set of students will meet for the duration of the academic term on a weekly basis, probably for 1 and 1/2 to 2-hour sessions. During that time, they will work through a set of problems that the instructor has provided, maybe with some input from you. At first, many of your participants will probably be looking to you for lots of information and even the answers to the problems, but after a couple of sessions, due to your reassurance and guidance, they will begin to rely most of the time on the group's fund of knowledge and expertise in problem solving. We say that, when a group is functioning well, an observer can watch the group for some time before determining which student is the official leader.

The workshop modules are developed by the course instructor; he or she may borrow or modify problems from the sample sets of questions previously developed by other faculty members using the workshop model and then develop some entirely new problems to match the needs of this particular class of students. At most workshop sites, students are not asked to solve the problems before the session, although they are expected to attend lectures conscientiously, read their texts, and complete their other homework for the course.

What? No answer key?

Also, in most programs, an answer key to the workshop problems is not provided, either to the students or to their leaders. As you might expect, this at first causes some grumbling among the troops, sometimes rather *loud* grumbling. The leaders themselves sometimes feel a bit at sea with this approach; they often tell us that they would feel more confident if they could face their students each week with the right answers tucked in their back pockets.

If we've stuck to our guns on this issue of answer keys, even after hearing these requests (well, complaints) for several years now, you might guess that we have strong motivation for doing so. No, it isn't because we want to "toughen up" students or put them through a hazing process of some kind. Instead, we know that answer keys can undermine the nature of what we're trying to create—that is, the students' confidence in solving tough problems, in working out strategies to develop their own solutions, and with finding answers that, in the group's opinion, hold up under scrutiny.

When you think about it, there are already plenty of "right answers" available in any course: in the back of the text, from old exams, in class, etc. But in real life (whatever that is!), we aren't provided with many verified right answers as we make our way through problems, both professional and personal. Engineers trying to develop new designs for more fuel efficient, cost effective cars can turn to previous work done by other engineers, and they can rely on the expertise and creativity of their own work groups. But there is no manual on their shelves that has the right design at the back of the book. Families facing decisions about how to raise and educate an autistic child have the experiences of others to draw from, but there is no official answer key for this very complex set of choices. Biologists setting up brand new labs may have an enormous amount of information about the equipment they could purchase, but—well, you get the idea.

So working without a net, so to speak, is much like the real world, but not much like the academic world most students have experienced to date. Many of the members of your group may firmly believe that it is your job—and that of the instructor—to provide "correct information," and that it is their job to memorize this to the best of their ability and then provide a reasonable facsimile of this data on the exams. We've included more information about this issue in the rest of the book, but we want you to emphasize the importance of students believing in themselves as real problem solvers.

The critical components

To sum up this section, we'd like to describe the six components of workshops that we understand to be critical to the success of the model. First, we believe that the workshop materials must be challenging and intended to encourage active learning, and they must work well in collaborative learning groups. It is essential as well that these problems are a good match for the students. We think of this as the Goldilocks principle: the problems can't be too soft or too hard. They need to be at just the right level.

Second, we believe that the job of facilitating workshops is a demanding one, so workshop leaders have a right to supportive supervision and to good training in the knowledge base of the course and in teaching and learning strategies for small groups.

Third, we know that it is essential for workshops to be integral to the curriculum and well coordinated with the spirit and structure of the course as a whole.

Fourth, we understand that the faculty teaching the course must be closely involved with the workshops and the leaders.

Fifth, it is important that organizational arrangements of the program, like the size of the group, location, time of day, noise level, and teaching resources, promote learning and good communication among the group members.

Finally, a workshop program needs the support of the institution's administration if it is to thrive long-term.

We hope that these opening pages have given you a sense of what the workshop model is all about and that you are feeling encouraged and excited about starting your first group. Remember that the readings in Part Two of this book are keyed to the chapters in Part One, so there is more information close at hand.

Now let's move on to one of our favorite subjects: you!

Bibliography for Chapter One

Hanson, D. (1997). *Process Workshop Homepage.* http://www.chem.sunysb.edu/hanson-foc/process.htm

Iadevaia, D. (1989). *A Study of the Relationship between Student Enrollment in First Year/Second Year Science Courses and Gender at Pima College*. ERIC document 310028.

Johnson, D. (1994). *The New Circles of Learning: Cooperation in the Classroom and School.* Alexandria, VA.: Association for Supervision and Curriculum Development.

Johnson, D., Johnson, R., & Smith, K. (1991). *Active Learning: Cooperation in the College Classroom*. Edina, MN: Interaction.

Martin, D. & Burmeister, S. (1996). Supplemental Instruction: An Interview with Deanna Martin. *Journal of Developmental Education*, *20,* 1, 22-24.

Reittinger, D. & Palmer, T. (1996). Lessons Learned from Using Supplemental Instruction: Adapting Instructional Models for Practical Application. *Research and Teaching in Developmental Education*, *13*, 1, 57-68.

Seymour, E. & Hewitt, N. (1997). *Talking about Leaving: Why Undergraduates Leave the Sciences.* Boulder: Westview.

Slavin, R. (1991). Group Rewards Make Groupwork Work. *Educational Leadership*, *48*, 5, 89-91.

Slavin, R. (1988). *Student Team Learning: An Overview and Practical Guide, 2d ed.* Washington, D.C.: National Education Association.

Swager, S., Campbell, J., & Orlowski, M. (1995). *An Analysis of Student Motivations for Withdrawal in a Community College*. ERIC document 387 000.

Treisman, U. (1992). Studying Students Studying Calculus: A Look at the Lives of Minority Mathematics Students in College. *College Mathematics Journal, 23*, 5, 362-372.

Chapter Two: The Role of the Workshop Leader

What is a leader?

If you have been a workshop participant, you have observed at least one model of a workshop leader. As you went through your semester as a workshop student, you probably found some traits and behaviors of your leader to be admirable and worth emulating—and others you'd just as soon discard. This is fine; all new leaders need to define and shape this role to suit their own styles and personalities.

However, there are a few "givens" that are worth discussing. It's probably obvious, for instance, that workshop leaders are not instructors. Although you have demonstrated your knowledge base in this course, you haven't had enough time to become an expert in the discipline. Instead, consider yourself an expert in *learning* this subject area, a real accomplishment in itself. (Please note that we have high hopes that many workshop leaders will pursue a teaching career. You are just the sort of people we want to be involved in the education of the next generation.)

But, getting back to your current role, you are also not expected to be teaching assistants. At many institutions, TAs are "junior experts" in the field and serve as "Answer" men and women in their work as recitation leaders, graders, and substitute lecturers in the course. Even with the best of intentions, many TAs find that the environment of their programs set them up as human encyclopedias, that is, students may consult and take notes from their TAs, but may not learn interactive, collaborative problem solving from them.

A harder distinction to make is the difference between a workshop leader and a friend, especially if you are close in age to your workshop students. Good workshop leaders are certainly friendly and supportive people, but, at least during the workshop sessions themselves, they are employees of their institutions. You've seen this in your best instructors; they are able to demonstrate a genuine personal concern for their students, and, during class time, they find ways to keep the focus on the students' academic needs. So, as a leader, you'll want to be congenial and more or less a peer, but not really a buddy.

A talented, experienced leader is much like a coach. If you think about it, good coaches occasionally demonstrate a particular move or strategy for you, but, most of the time, they sit on the sidelines and call out timely advice and encouragement while *you* run down the track, shoot for the basket, or kick the ball. People struggling with difficult material need to know that somebody believes in their ability to succeed. A few sincere remarks sprinkled through the session, like "I like the way you set up that problem" or "I can see that you are getting the hang of this" can do much to keep a group moving.

A good leader is also a troubleshooter. Smoothly functioning groups seldom happen just by luck; they require careful nurturing over time. A good leader spots problems with group dynamics and with individual students, and then helps the group find its way through these thickets (more on this topic later).

Probably most importantly, a good leader serves as a role model. You are living, breathing proof that students can succeed in this subject, and while you are undoubtedly not the average person on the street, you are not some sort of mutant genius either. In terms of motivation, workshop participants benefit greatly from their time with leaders, their "near peers." Just by being there, you provide assurance that normal human beings can conquer this mountain of material—and maybe even enjoy it.

Ethics

The ethics of being a workshop leader deserve some attention, too. As in all professional positions, being a workshop leader carries with it a set of ethical responsibilities. You would figure these out in the course of a few sessions anyway, but we would like to provide a few ideas to start you thinking.

The first part of these ethical guidelines concerns what the students and your instructor anticipate from you. Naturally, they should be able to count on you to show up for your workshop session on time and on your reliable preparation for the module of the day. There is no surer way to shoot down a workshop group than to have a leader who doesn't follow through. This sounds obvious, but, given the demands of the rest of your life, we figure that living up to these professional expectations will take some planning on your part. After all, you have your own classes to attend to, maybe another job to handle, perhaps some family responsibilities to meet, and a social life to enjoy.

So now is the time to think through how to manage things so that the duties of being a workshop leader fit into your life. Because you have proven success as a student, we believe that you probably are pretty savvy about organizing your time, but, since being a leader may be a new domain for you, we have included a section on time management in Part Two of this book.

And don't forget how much help your fellow leaders can be. Get together often with some of your colleagues. Collaborating with them is your best way to check your own understanding of the workshop modules; it serves as a role model for the type of work you want your own students to do, and, best of all, it offers you a great way to connect with some of the very best students at your institution.

After considering what the program can expect from you, it's time to think about what you can expect of yourself. One matter is determining how much time and effort you should dedicate to this work. We want your heartfelt commitment and participation, but not at the expense of your own well-being. Your program directors can approximate how much time their effective leaders devote to their workshops. A common estimate is that being a good leader takes a weekly commitment of about six hours: two hours for the workshop session, another hour for the training meetings, another hour or so to review the problems with fellow leaders, and one or two more for the reading and preparation for the training sessions.

While the actual number of hours needed varies from site to site, it is always the case that leaders need to establish boundaries for themselves. It's easy for leaders to be buttonholed by panicky students in the cafeteria, in the library, and, heaven forfend, in the dorm at 1 a.m. We would just as soon you *not* make yourself universally available; this is a good recipe for burnout. You may occasionally want to set up an extra review session or two, like right before a midterm, but let this be according to your own plans and schedule.

Another part of leader ethics is being alert to issues outside the area of your own expertise. This starts, of course, with the subject matter itself. Don't let your students make you feel responsible for knowing all; when they ask you questions you can't answer, a whole range of possible responses can get you off the hot seat. Here are a few:

1. "That's a great question. Let's take a look in the text to see what the author has to say on this issue."

2. "That's a great question, but it's a bit beyond me. I'll check with the instructor tomorrow, and I'll email you all the answer."

3. "That's a great question. Who has a start toward answering it?" (We like this one best.)

Another way you can get pulled into something beyond your own professional experience is when a student comes to you with a personal matter, like an impending divorce in the family, information about an eating disorder, an abusive relationship, or another issue of this sort. Leaders hear about these things fairly frequently, not a big surprise, since they are competent, caring people working in a small group setting. Remember that you aren't responsible for solving students' life crises. We *do* want you to be attentive, listen long enough to students' concerns to understand what the issues are, and then be adept at referring your students to appropriate on- and off-campus resources.

Finally, let's turn to a sticky part of boundary setting: dealing with personal relationships between you and individual group members. These connections may come as part of the package, i.e., you may already be friends with someone who is placed in your group. Often, this works itself out naturally, but, on occasion, either the leader or the student in the workshop feels uneasy with this change in role. When this happens, we've discovered that a private conversation, initiated by the leader, often smoothes things out. Here's how it might go *outside the workshop setting:*

Leader:	Have you felt sort of weird during our workshop sessions?
Student:	Yeah, it gets a little uncomfortable sometimes.
Leader:	Guess we're mostly used to being friends, and it feels a little strange with me kind of "in charge."
Student:	Well, I understand, but mostly I just don't want you to think I'm an idiot if I need to ask stupid questions.
Leader:	Hey, just last year I was in the same class, and I thought my questions were maybe dumb, too. But my leader told me that everybody figured their questions were stupid, and really none of them turned out to be dumb at all. I promise you can ask me anything, as long as you promise to speak up whenever there's something you're unsure about.
Student:	Ok, this makes me feel a lot better. But don't you dare start bossing me around, all right?
Leader:	Ok, sounds good.

Even more confusing are the romantic attachments that may develop between a student and the leader during the course of the academic term. It's not hard to see how this can happen; after all, you're students at the same institution, and you have a lot in common. From the workshop participants' point of view, you can be a very attractive person: you are there in a friendly, helpful role, in a situation in which the student may feel scared and insecure. You spend quite a bit a time together, and, by nature, those students who are selected to be leaders tend to have strong people skills. So it's easy to understand why a student in your group may be especially drawn to you. And it isn't outside the realm of possibility that you yourself may be attracted to a particular student. Someone in the group may have that certain something that makes your workshop times together seem extra special.

So then what? Unless the two of you would be prepared to run away on the spot and join the circus together, you would have to deal with this smack in the middle of regular life—and those weekly workshop sessions with the other students in attendance as well.

Over the course of our project, we've posed this hypothetical situation to quite a few leaders. They tell us that allowing the student to follow through on his or her feelings or letting leaders take action on their own could result in all sorts of problems. Even if the feelings are mutual, other students in the group are bound to catch on and feel excluded from this cozy arrangement. And dealing with unwanted advances can feel like pressure, and even sexual harassment.

The bottom line from our previous leaders goes like this: if the attraction isn't mutual, kindly, but plainly, tell the student that you need to keep a professional relationship in place. If the feelings *are* coming from both of you, remember that an academic term is a very short span of time. It might be hard to stick to boundaries for awhile, but if there is anything good between you, it will last until your official duties with that workshop group are concluded.

Here's one of our favorite (and true) stories about this kind of thing, not really from a workshop setting, but close enough:

> *A few years back, a graduate student was teaching one of her very first courses. She had noticed that one of the older students seemed to be paying more attention to her than she expected, but he didn't say anything, so she just went about her business as an instructor.*
>
> *After the last day of the term, he came by her office. He asked first if she had graded his last work. She said yes. He next asked if she had turned in all of the final grades. She said yes. He then asked her if he was right, that for both of them the course was concluded in every way. She said yes again.*
>
> *Then he took a big breath and asked if she would go out with him. She said yes then, and one more time when he asked her to marry him a year later.*
>
> *Last we heard, they were doing just fine.*

So, if this happens to you, we hope you live happily ever after.

Chapter Three: Getting a Group Started, Keeping It Going

Preparation

Even though many of us involved in the workshop project have been teaching for a good long time, we can still get the willies when we face a new roomful of students. What will the "personality" of this group be like? Are they enthusiastic about being there? Will anyone in the group be a particular challenge? And, to be honest, we wonder, will this group like me? So we understand that you, perhaps with your first group of students, are likely to feel a few qualms yourself. This chapter is designed to help you through those early sessions until the group is ready to take on more responsibility for maintaining itself.

There is nothing that will help as much as good preparation. For the very first session, plan on spending twice as much time getting ready as you will need later in the term. Start out with the content itself. Re-read the chapter(s) keyed to the first workshop module. This is fun in itself since you'll notice so much more this time than when you read this material as a new student in the course. In terms of the workshop, you'll find that no textbook is perfect. Now you are in a good position to see where the explanations are too complex for a beginning student and where the author has skipped over some necessary steps or failed to define new terminology in a meaningful way. Make note of these places, not with the goal of mocking the author (it's amazingly difficult to write good beginning chapters to introductory texts), but rather to know in advance where your students may be confused. You want your group to look at the text as a helpful resource, even if it has some weak spots. A good way to do this is to prepare a quick summary sheet of the current reading to collect the main points for your students and fill in the blanks. Students appreciate their leaders' efforts in pulling together cheat sheets like the these. Not only do these summaries help students see the forest for the trees, but they also encourage students to prepare these sheets on their own for this course and for their other classes. Since most college students feel like they are drowning in information, anything that helps to organize and clarify matters is a plus.

It's also crucial that you spend preparation time with the workshop modules. Even if these are the same problems you worked your way through last year as a member of a workshop, you still need to review them in detail. Here are a few things to keep in mind while doing so:

1. *What will the students' reactions be when they read the problems for the first time?* We're guessing that a couple of students in each group will be legitimately confident about working through the problems because they have some solid background and preparation in this area. One or two others may be overconfident; sometimes those who have had a little exposure to the material in a previous course overestimate what they are ready to handle now. But most of the group will feel confused, even a little panicky. They may be used to factual question-

and-answer worksheets in their previous courses, so the complex problem solving required by a workshop module can seem daunting.

2. *What terminology needs decoding?* Remember that it takes many exposures to new words to learn them. So even if students have been introduced to these new terms in the text or during lecture, it isn't likely that they will have internalized their meanings at this point.

3. *Where are the clues in each problem that will help students get started?* You don't want to walk them through the whole problem-solving procedure, but you should be prepared to give them a handle on the questions during the first couple of sessions.

4. *Where are the bear traps?* What is likely to lead them astray while they are trying to get through particular problems? Your job is *not* to prevent them from doing so—following wrong leads is a part of finding the right direction. But you will want to know where they are likely to become bamboozled, and to have a couple of strategies or clues in your pocket to keep them from spending the whole two hours trapped in a snare.

5. *How will you confirm your own understanding of the solutions?* As mentioned before, your fellow leaders are your best connections in this program, so we strongly urge you to meet weekly with a couple of your counterparts, just to review your own preparation and compare notes. Many leaders have told us that this is one of the most rewarding elements of the workshop program.

And don't forget the instructor. Again, he or she doesn't expect you to be a professional chemist, biologist, mathematician, or whatever. These folks are eager to help you feel prepared for your workshop sessions, so please don't feel embarrassed about a need to review some of the basics. Remember that they have been through this material many, many times, and they know you haven't had this opportunity yet.

6. *Are you ready to create a comfortable working environment in your group?* Make sure you have all your paperwork (e.g., attendance lists, workshop problems) at the ready. Review the list of all the students in your workshop. If the group is not diverse (e.g., no males), talk to the instructor about the distribution of gender. Have an icebreaker activity planned (more on this later) to promote positive group dynamics.

Preparation also includes surveying the physical space you have to work with. As instructors, we inspect any new classroom to which we are assigned, just to forecast how this space is likely to match our teaching styles. You'll want to do the same, to find out how well your room will fit the needs of a workshop session. An ideal workshop room is fairly small, so people don't feel lost in a cavernous space. It has comfortable chairs and several small tables that are all easily moved into a square or circular configuration. There is plenty of black- or white-board space, with lots of chalk, markers, and erasers at hand. The room itself is in a quiet location and has good acoustics, so no one needs to shout to be heard.

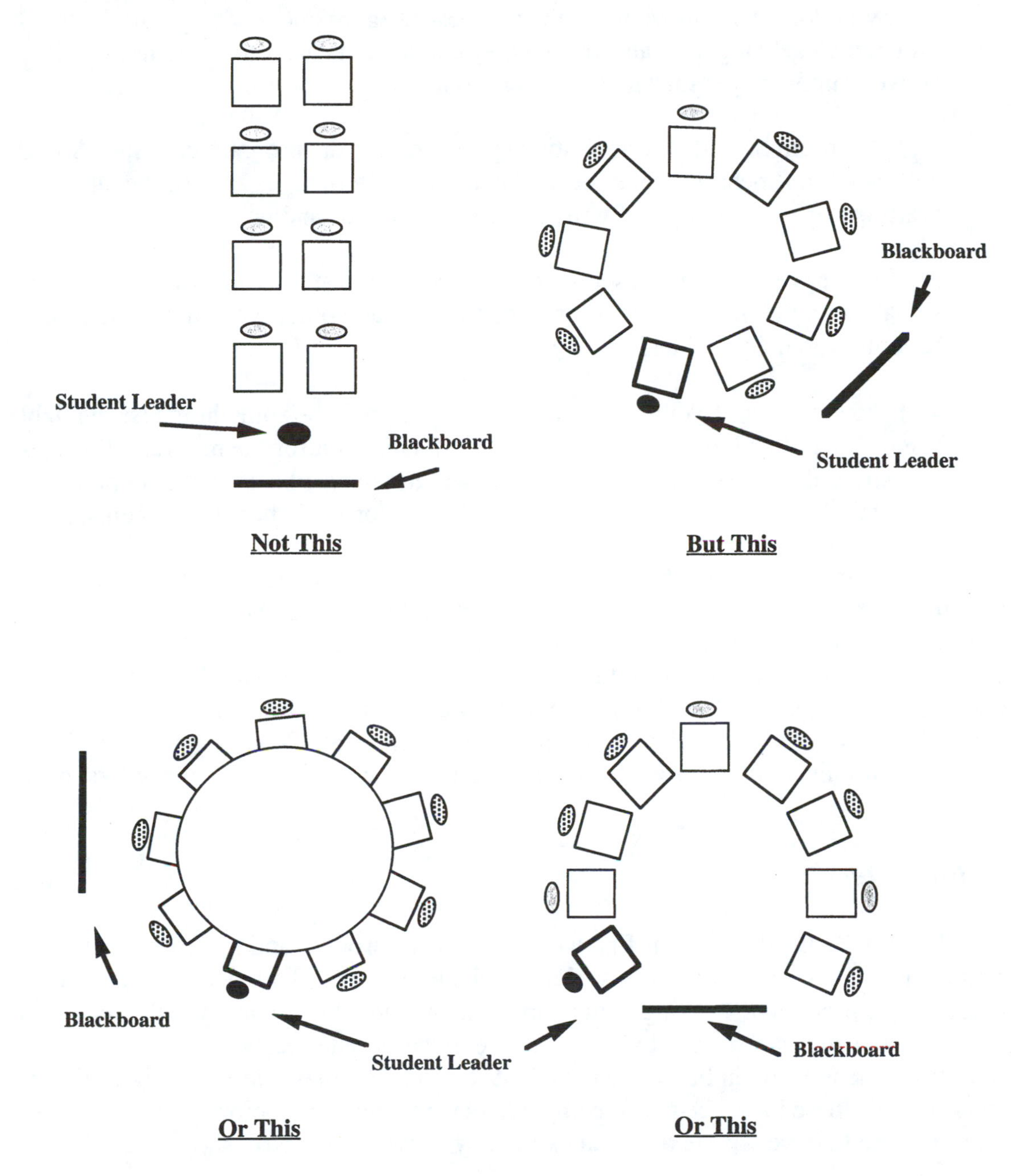

We hope that this is an accurate description of the room you've been assigned, but we know that it's often necessary to work with less than perfect space. "Real estate" is frequently at a premium on many campuses, so you might find yourself looking at a room that doesn't correspond well to the description above.

Here are a few suggestions about what to do if your space seems less than ideal:

1. If you don't have movable chairs and desks, figure out how to get students grouped closely together anyway, so they can look at each other and hold real conversations, rather than all staring ahead at you!

2. If the room has little or no writing space, contact the program directors ASAP and secure a flipchart or an easel, and markers. Alternatives are big sheets of drafting paper or a couple of write on-wipe off boards and pens.

3. If you have to share this space with another group at the same time, work out signals with the other leader beforehand to alert each other when the noise level becomes too much to handle.

4. If you really hate your room, speak up. There may be something else available somewhere. In a large workshop program, it's difficult for the program directors to inspect all the assigned rooms, so they will appreciate hearing about conditions in certain locations that make them poor choices for workshop assignments.

Also, don't be afraid to be a room detective. We've been clued in to many good places for workshops on our campuses by our leaders themselves. For example, some dorm study rooms that we didn't know about before have been really terrific places for workshops, even if the leaders need to bring in flipcharts for a place to write. These lounges may have just the right furniture, and they can be a nice break for students who get tired of spending so much time in the classroom each week. You might know of other little conference rooms around campus that would be good choices for workshops.

The First Meeting

Eventually the day of your first workshop comes around, and it is time to meet your students. As we have said, please don't feel silly if you feel a few butterflies. It's just part of the process of working with people who are new to you, and it will pass, trust us. We would suggest arriving at your room five to ten minutes early to make sure you've had time to settle in before your students arrive. You may want to bring chalk or a marker, in case these have been pilfered by others using the room before you. And you'll want to be sure to have your text and an extra copy or two of the workshop modules.

As the students file in, introduce yourself to each of them and give them a little something to do, just to break the tension of being in a new group. You need them to get to know each other anyway, so you could give them all pieces of typing paper and ask them to make table signs with their names on them. This is a little kindergarten-like, but it really does help people remember new names.

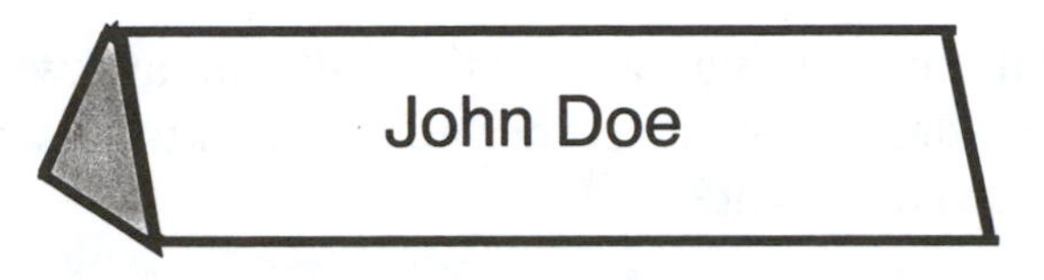

If they start to spread out around the room, coax them into moving so that they can see one another. Students are so accustomed to sitting in rows and looking to the front of the room that rearranging them may take a little nudging, but don't give up.

Once it looks like you have a quorum, set aside 15 minutes for this first meeting to allow students to interview each other. Each student should select someone they don't already know well to interview and then introduce to the group. The goal here is really just to get every single one of them to say *something* in the group right away. These sessions work only if everyone is a part of the process, so you don't want anyone to take on the role of the group wallflower. During this first meeting, working on a positive group environment is your main job.

It's also crucial that you spend a few minutes describing yourself and your job as a workshop leader. You'll need to do this in a way that fits your own style and personality, of course, but you'll want to get the idea across that you aren't the junior instructor or the human answer key and all that. Instead, establish yourself as somebody who is a step ahead of them, and who knows where the quicksand is in this material. Remind them that you plowed your way through similar work recently, and that now you are eager to help this group do the same. You don't need to mention your GPA, but you will want to assure the group that you found the course do-able.

You should also describe the group process very briefly. You probably don't want to spend much time doing this, or you'll run the risk of lecturing at them about how you don't want to lecture! But you can take a minute or two to let them know that the real power of the workshop model is right there in the group itself, and that, among them all, they will be able to find good problem-solving approaches and good answers to the workshop problems.

Since the "bonding thing" is crucial to the success of the format as a whole, you may choose to spend a little time next with some ice breakers. One idea is to have everyone to sit in a circle. The first person then states his or her name and two personal descriptions. The next person then introduces the first person, and states these two things again. The game proceeds around the circle, with each person introducing all who came before and their two descriptors (Forbess-Green, 1983). A second strategy is to divide the group into teams of six or fewer. Ask each team to design and present a 30-second "commercial" for their project, in this case, perhaps some concept or component of the course (Silberman & Lawn, 1995).

Don't feel bad if these beginning activities feel a bit awkward and artificial. It takes time for people to warm up to each other, to you, and to this new environment, but you've got to start somewhere, right?

After you've helped your students reach a first name basis with each other, it's time to get down to the work itself. You want everyone in the group to conclude this first session feeling that something substantial has been accomplished, so it's acceptable to be a bit directive about getting everyone focused on the first module. If you're not used to getting people to work, here are a few phrases you can try:

Well, what do you guys think of the problems in Module One?

Ok, now that we know each other a little bit, let's take a look at those problems the instructor handed you for this week.

Ok, enough fun and games everyone! Let's see what we've got to work with in Problem One.

Despite your enthusiastic tone of voice, chances are pretty good that you'll be met by some blank faces or even moans and groans as you try to get the group focused. No sense in getting defensive about this; solving these problems is pretty hard work, and it's just human nature to try to avoid it. So let any resistance you receive slide right off and help them get going on the first problem anyway.

One of the students has to take the bait at this point. Being the first student to speak up about a workshop problem is a fairly precarious role; you'll want to make this as easy as possible for somebody to start talking. An almost painless way to do this is to ask someone to read the first question out loud, promising this person that he or she will *not* be expected to start answering the question right away. We say that this is *almost* painless because sometimes there are words within the question that some of the students don't feel comfortable about pronouncing yet. This brings us to an essential principle of working with other people:

Many people would rather fail later than look stupid in front of others at the moment.

As a workshop leader, you should keep this idea in mind at all times. It is one of the key reasons why lectures alone aren't a very effective instructional system: there are far too many ways to look dumb in front of your peers and the instructor if you speak up.

The goal is for the workshops to be a place in which a few risks can be taken and mistakes can be made without feeling terrible. You can help right away with this "read the question out loud" plan by using the difficult words in the question in your own request:

Could I get somebody to read this problem about stoichiometry aloud for the group, just to get it on the table?

In the next section, we're going to look at the differences between psilophyta and lycophyta. Would somebody read this question for us?

If other pronunciation errors pop up during this reading aloud or during the ensuing discussion, you probably don't want to point it out directly. Instead, you can use the word yourself correctly a minute or two later in the conversation; the students will catch on.

After the problem is in front of the group, it's time to figure out some approaches to it. They may look nervously at you with big puppy eyes, but resist the temptation to start teaching, no matter how sorry you may be for them. Instead, get them to start picking apart the problem. Here are some ideas:

Any terms here that we should define before we go on?

Well, what do you figure this problem is really asking us to do?

H'mm—this is a challenge, isn't it? What information here seems the most essential to you right now?

What do you remember from the lecture or the text about this topic?

And maybe most importantly: *I have a feeling it's going to take all of us to solve this one.*

Another important part of starting up the group is making people feel ok about going up to the board or the flipchart. Nobody wants to look like a fool standing up there; so make sure that your volunteers (or your draftees) know that they are being asked to serve at first only as a scribe for the group. Their job is just to write down what the group calls out, step by step. This way, you can get your students used to standing up with a piece of chalk or a marker in hand without making them feel too exposed. An added incentive: the student who has gone up to the board and written something gets to pick the next person to take the chalk. (We're not sure *why* this works, but it does, and it takes the pressure off the leader.)

Also, a little amiable competition can be a good thing, even in a program where cooperation is a major goal. So try dividing your students into pairs or triads and set up a friendly contest between the teams. This is a good way to get the students talking to at least one other group member, and it keeps them from looking at you every step of the way.

During this first meeting you are sure to notice differences among your students in personality, in their approaches to the workshop format, and in their experience with the material. A quick note here: be careful *not* to make judgments about the comparative braininess of your students because it's easy to be confused at the beginning. For example, many people associate an outgoing attitude with intelligence and a quiet approach with the opposite. But what you're probably noting at first are differences in personality, not in cognitive ability.

But these variations in personality and approach mean a lot. Consider the following vignettes and predict how the person in each story will be likely to fare in a workshop over the course of the semester. Will this person help the group accomplish its goals? If so, why? If not, is there some way that you, as the group leader, could redirect this person towards a more positive experience in the group?

Steve

Steve came into your group on the first day right on time and soon had his notebook and pencils at the ready. He was at work on the problems before you even started off the group, so you were convinced you had a go-getter here. But, when it came time for the group to work together, he wouldn't speak up or contribute at all. Several times you glanced at his paper and saw he was on the right track, but every time you encouraged him to share something with the group or work with another student, he looked away or mumbled something about not knowing the answer.

Louise

Louise marched into the group the first day and announced to the group: "I don't know why we all have to be here. I already worked through every one of the problems the professor gave us." When you looked at her paper, though, you could spot several places where she had the wrong idea. Nonetheless, she clearly seemed to feel sorry for the other group members and started to tell them how to do the problems. You are a little worried, first, because her own understanding appears hazy, and second, because she was leading the rest of the group down the wrong path, both in terms of content and attitude.

Lee

Lee came late on the first day of the workshop and sat off to the side. He reluctantly joined the group in the center of the room when you encouraged him to move. He had forgotten his workshop problems, so you had to give him your spare copy. The other students got right down to work, but you had to coax him into starting the first problem. When you asked for volunteers to come up to the board, Lee wouldn't even make eye contact with you. You hoped to touch base with him at the end of the session, but he was out the door before you had a chance to catch up with him.

Maria

Maria came into the room on the first day, sat down between two quiet students, introduced herself and said, "Wow, I just looked over these problems—they look pretty challenging to me! I'm so glad I don't have to face this stuff all on my own." After you started the group on the first problem, Maria turned to the person sitting on her left and said, "H'mm, what do you think? Am I on the right track here?" Soon the two of them were comparing their strategies and holding a friendly argument about the answer.

These four students are all made up, but the group issues represented here are very real. With sincere apologies to all the researchers who have outlined the intricacies of group dynamics, we like to start with a very simple way of looking at how different attitudes mesh with the workshop model. Consider for a moment the following matrix:

Attitude	Effect on Workshop	Effect on Workshop
outgoing	positive	negative
quiet	positive	negative

With a model like this in mind, you may decide that the students described above would fit into the grid as follows:

Attitude	Effect on Workshop	Effect on Workshop
outgoing	positive—MARIA	negative—LOUISE
quiet	positive—STEVE	negative—LEE

You may have a different take on these students, but in any case, it's easy to see that both outgoing and quiet students can be competent in the subject at hand. We've found that quiet students like Steve often become more confident once they feel assured that they won't be put on the spot during the workshop sessions.

It's also important *not* to give up on students like Louise and Lee. Instead of deciding that they are just "that way," try for a moment to look at things from varied points of view. Sometimes students who act a little bigheaded, like our Louise here, are feeling unsure about themselves, so this attitude serves as camouflage. Or maybe they

mistakenly believe that their Advanced Placement course in high school really did cover everything in a college course (the first couple of weeks of the semester may lull them into a false sense of security). If they've started out acting a bit obnoxious, it may be hard to feel sympathy for them when the inevitable truth (that they really *don't* know it all yet) makes itself known, but serving as a safety net is part of the job of being a workshop leader. And you may find that the Louises of this world can turn into Marias, with the proper nurturing.

As for someone like Lee, be careful not to jump to conclusions. He may be behaving like he just doesn't care, and who knows? Maybe he doesn't. But perhaps there is something else going on, a family crisis, big money problems, a broken engagement—or being too nervous about the course to get involved yet (you can't fail if you don't try, goes this logic). We've found that the workshop model is essential for the Lees of this world. In a lecture-only format, especially in large classes, they tend to disappear in the crowd. But *you* are certain to notice that something is happening with him, and you, along with the rest of the group, may be the ones to make a good connection.

One last note about these students: it's easy to let the squeaky wheels dominate your attention. But don't forget about Maria. She seems to be getting off to a good start, but she still needs you and the group over the course of the term. In fact, you may want to take special care of the development of Maria's competence with the material and with the dynamics of the group; she might make a great leader next year.

More Tips for Leaders

1. *Identify the needs of each of your group members.* In many groups, there is considerable variation in students' abilities, with some being substantially more adept with the material than others. This can seem alarming at first, but the research says that groups with uneven abilities actually have advantages over those with more even distribution of skill (Johnson & Johnson, 1992). So if you have a "mismatched" group, don't worry about it. Instead, make good use of the more advanced students' abilities; let them teach the others some of the time. See the reading selection by Epstein for more ideas on this topic.

2. *Throughout the semester, plan for adequate time to prepare for your workshop.* Make sure you have figured out the problems and that you have a good idea about where your students will stumble, keeping in mind the different abilities and motivations among your group members. It helps a great deal to put together some notes or a gloss for each week's problems; that way you have something to refer to if you get stuck during the workshop session itself.

3. *Use your own experience as a guide.* What helped you understand the difficult concepts in this course? If these strategies worked for you, they may help some members of your group, too.

4. *Try the one-minute paper.* This idea, developed at Harvard, is very simple, but it pays off big (Light, 1990). Near the end of the workshop, ask students to jot down responses to two questions: "What was the main point you learned today?" and "What was your biggest unanswered question?" As they leave, they drop their anonymous papers in a box near the door. These responses provide a window for you about what is making sense for your group and what isn't. And *your* response to their feedback makes for a great introduction in the next workshop—they get their questions answered, and you can build some continuity of ideas across time.

5. *Make use of the faculty.* Workshop leaders sometimes treat their job like one big test (i.e., that they have to prove themselves somehow during the term by knowing everything). But this program isn't a test at all. So it's fair—and very desirable—for you to keep in close contact with the instructor of the course. Ask questions. Provide feedback about the workshop problems. Keep in touch.

Bibliography for Chapter Three

Forbess-Green, S. (1983). *The Encyclopedia of Icebreakers: Structured Activities that Warm-up, Motivate, Challenge, Acquaint and Energize.* San Francisco: Jossey-Bass.

Johnson, D. & Johnson, R. (1992). What to Say to Advocates for the Gifted. *Educational Leadership, 50*, 2, 44-47.

Light, R. (1990). *Explorations with Students and Faculty about Teaching, Learning, and Student Life, First Report.* Cambridge, MA: Harvard.

Silberman, M. & Lawson, K. (1995). *101 Ways to Make Training Active.* San Francisco: Pfeiffer / Jossey-Bass.

Chapter Four: Multiple Intelligences, Learning Styles. and the Workshop

Do you know your IQ score? If so, this number may make you feel pretty smart, kind of ho-hum, or maybe even a little embarrassed and put off. In any case, your number and your resulting reaction to it probably deserve reexamination, since the notion of global intelligence has recently been under attack. Many now consider this model to be a fiction, an artifact created by the methods used to measure cognitive ability.

The study of intelligence caught fire in the 1800s, with the development of the belief in psychology as a science. James Cattell is credited with coining the term "mental test" a little over 100 years ago, but it was Alfred Binet and his associates who developed the first widely adapted intelligence test in the early years of this century. His goal was to understand why some children fail in school. His scale, therefore, was not intended to measure intelligence of the population at large, but to predict which children would benefit from school instruction. Later, Henry Goddard imported Binet's test to the United States, and it was his student, Louis Terman, who adapted this instrument while working at Stanford University, thus giving the world the first Stanford-Binet test (Thorndike, 1997).

The notion of a general intelligence quotient caught on in educational circles, and for several generations now, we have often acted as though overall cognitive ability is captured by a few numbers. It's true that scores on the WISC (Wechsler Intelligence Scale for Children) and the WAIS (Wechsler Adult Intelligence Scale) correlate with certain kinds of academic success in traditional courses, but this is not surprising, since Binet and others originally developed these tests to sort for those who would do well in traditional school settings.

But common sense tells us that people can be smart in different ways. With a little reflection, you can think of important forms of knowing the world that aren't measured very well by the standard measures of intelligence—and some types that aren't tapped at all. For example, the ability to see how things are put together is assessed to a modest degree by the usual IQ tests, but not very completely. The ability to learn musically isn't evaluated at all, and neither is our ability to understand and work with others or to make sense of our own internal motivations and thought processes. So the "ways of knowing" that a good auto mechanic or dancer or musician bring to the world aren't reflected in standard IQ scores, and neither is the crucial ability to understand and work with the intra- and interpersonal components of our lives.

This seems obvious to many of us, and, in fact, the first debates about measuring mental abilities included considerable discussion on the issue of multiple intelligences. Early on, E. L. Thorndike understood how complex a good test of cognitive ability would have to be: "The primary fact is that intelligence is not one thing but many. The abilities measured by a speed test with language and mathematics are not identical with, or even very similar to, those measured by a test with pictures and less exacting in speed" (1920, p. 287).

But the idea of general intelligence held the day for several generations, maybe because there is something convenient about a neat and relatively easy way to label people. Recently, however, there has been a resurgence of interest in the notion of multiple intelligences. According to this theory, these different abilities are regarded not just as talents or proclivities, but rather as different ways of taking in and demonstrating knowledge of the world. There currently exists a number of ways of categorizing these learning differences (Krause, 1998; Dunn, Dunn, & Price, 1985), but a useful schema for workshop leaders is the one proposed by Howard Gardner (1993; Chen & Gardner, 1997). Here are the types of intelligences posited in his theory:

linguistic — People with this type of intelligence access and display knowledge well by listening, speaking, reading, and writing. As you might guess, these students are often advantaged in traditional teaching formats.

musical — Students with strong intelligence in this category are good at understanding patterns in pitch and rhythm, which sometimes can help them see other sorts of patterns, like those in math. When children are taught new information, it is often cast into a song; we all can sing our ABCs. Music can help us remember other things, too; that's why we hear so many jingles on TV. Sometimes we can use this approach with more academic information, too; if you hang around biochemistry students, for example, you may hear them singing the Kreb's cycle set to an old British pub song!

logical/mathematical — Students who do well at quantitative thinking—which also includes certain types of sequential learning—also tend to function well in much of the conventional academic world.

spatial — You remember those aptitude test questions with the various geometric shapes and the question "Which of the following matches the target shape?" Or, as a harder version, "Which of these shapes is a rotation of the target shape?" If you are good at answering questions of this type, you probably have considerable spatial intelligence. This is of real advantage when thinking about molecules, anatomical structures, and other components of science learning.

kinesthetic/bodily	Students who excel in this type of learning are good at gathering and demonstrating knowledge through large and small muscle activity. Hands-on learning is very much part of good elementary curricula, but tends to receive less and less focus as the grades advance. By college, there often is very little opportunity to learn this way, except in special environments like labs and studio arts courses.
intrapersonal	This type of intelligence is not merely self absorption. It is, instead, the ability to understand and make good use of our own thoughts, feelings, and motivations. This category is often thought to be of most value to those involved in highly introspective pursuits, like writing poetry or studying philosophy. However, all students who know themselves—their motivations, their weak spots, their ethical perspectives—are in a good position to make the most of their college experiences. They know how they tick, so to speak. And in nearly any leadership capacity, like being a project director, a lab manager, or a workshop leader, those who have the ability to understand *themselves* honestly are well positioned to tap the next intelligence described below.
interpersonal	Students who are good at interpersonal learning flourish in situations in which they are allowed to talk through their ideas and hear the opinions of others. These students do very nicely in seminars, but find the lecture format to be less conducive to learning. Current research on student success in general tells us that this way of knowing is very important for many students' success, especially in math and science (see, for example, Light, 1992).

Gardner's theory of multiple intelligences is under continuing development. Recently, he has reviewed evidence for the existence of additional intelligences (Gardner, 1999):

naturalist	This intelligence, Gardner suggests, incorporates the ability to build and use taxonomies in the environment, i.e., systems of recognizing and classifying members of a species or category. Someone who is strong in this intelligence can not only distinguish among individuals within a species or group, but also can understand relationships among species. An entomologist studying butterflies in the rain forest is making use of his/her

naturalist intelligence. A strength in this area can be an advantage for students in the sciences, especially those in inquiry- or field-based courses.

existential — Gardner acknowledges that the ability to consider the "existential features of the human condition" has been valued in many cultures throughout history. He is not sure, however, if the capacity to "locate oneself" in regard to the meaning of life and death qualifies as a separate intelligence. He says "I find the phenomenon perplexing enough and the distance from the other intelligences vast enough to dictate prudence—at least for now" (1999, p. 66).

Thinking about these differences in the ways we know the world can help you understand the members of your own Workshop better. You can learn more about the application of these ideas to the study of science by reading the article by Felder in Part Two of this book.

Different intelligences, defined by Gardner as "the ability to solve problems or create products that are valued within one or more settings" (1999, p. 33) are not the same thing as learning styles or preferences. A learning style is the way we like to put our different intelligences to use. It's fun to find out about our different learning preferences; you can do so by completing a learning style inventory like the one found in McWhorter (1998) and many other study skills texts. You're bound to discover some unexpected ways in which your approaches to learning vary from those of your fellow leaders, even though you all have demonstrated academic success in a similar arena. For instance, we administered a learning styles instrument to a group of humanities faculty members—all with similar backgrounds, all with similar teaching and research responsibilities—and they were amazed to see how very different their learning profiles turned out to be. There can be many routes to the same goal.

Let's assume for the moment that these differences among us are genuine. So what? Why should this matter to a workshop leader? The answer is pretty straightforward. If you *aren't* tuned into these different intelligences and learning styles, you will tend to present information and respond to input in the ways that work best for you personally. It's just human nature; we explain things in the fashion that makes the most sense to us. This isn't altogether a bad thing; these methods work for us, right? However, if this is *all* we do, we advantage the students who happen to be similar to us, and we inadvertently disadvantage those who are less like us. Since our workshop goal isn't to replicate our own clones, we are responsible for making our learning environments more pedagogically diverse.

Let's look at a simple example from everyday life. Say, for instance, that you want to give directions to a party at your house. If you are the type of person who learns best through words, you might say something like "Go north on Park for two miles, turn right on Elm, stay on Elm for four blocks, take another right on Turner, stay on Turner until you come to the intersection of Turner and Brooks, take a left on Brooks. We're the third house on the left."

For some folks, these instructions will work perfectly; they will show up at your party right on time. For your party guests who like visual cues, adding some information can help, like "When you come to Elm, you'll notice a large granite building on your left," or "Our house is a blue colonial with white trim." For people who are really spatial learners, all these words just muddle things up. A map might work great for them.

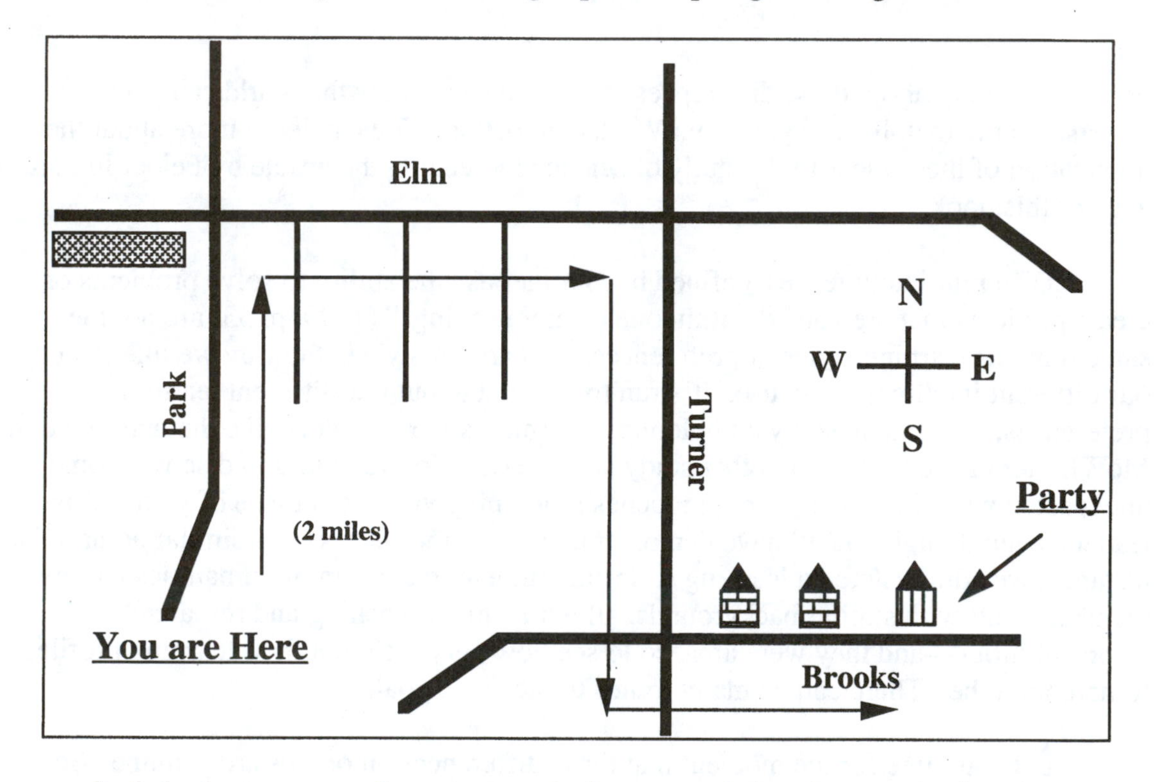

Let's look at a more relevant example from a workshop in organic chemistry:

All of the following reactions occur by analogous mechanisms. Write a single, general mechanism that explains all of these reactions. Make a table listing all the electrophiles and the nucleophiles, reaction by reaction, for the first mechanistic step and also for the second mechanistic step.

$RCH{=}CH_2 + CL_2 \rightarrow RCHClCH_2Cl$

$RCH{=}CH_2 + CL_2$ (in H_20) $\rightarrow RCH(OH)CH_2Cl$

$RCH{=}CH_2 + Br_2 \rightarrow RCHBrCH_2Br$

$RCH{=}CH_2 + Br_2 \text{ (in } CH_3OH) \rightarrow RCH(OCH_3)CH_2Br$

$RCH{=}CH_2 + HCl \rightarrow RCHClCH_3$

$RCH{=}CH_2 + HBr \rightarrow RCHBrCH_3$

$RCH{=}CH_2 + H_2SO_4 \rightarrow RCHCH_3$ (with O-SO_4H attached to the CH carbon)

$RCH{=}CH_2 + H_2SO_4 \text{ (in } H_20) \rightarrow RCH(OH)CH_3$

$RCH{=}CH_2 + H_2SO_4 \text{ (in } CH_30H) \rightarrow RCH(OCH_3)CH_3$
$RCH{=}CH_2 + Hg(O_2CCH_3)_2 \text{ (in } H_20) \rightarrow RCH(OH)CH_2\text{-}Hg0_2CCH_3$

$RCH{=}CH_2 + Hg(O_2CCF_3)_2 \text{ (in } CH_30H) \rightarrow RCH(OH_3)CH_2\text{-}Hg0_2CCF_3$

The problem presents twelve different chemical reactions and asks the student to find the common explanation (mechanism) that transforms the reactions to twelve specific examples of a more general concept. The big idea here is that the mechanism connects the twelve reactions and makes it much easier to understand and learn them.

At the most fundamental level, this problem requires the ability to find the pattern in a complex set of information. *Logical* thinking will be very helpful. In practice, the problem directs the students to this kind of analysis by asking them to make a table. A table is a way of identifying and displaying a repeating logical pattern. Pattern recognition skills are also characteristics of *musical* thinkers; they will find it natural to construct the table. The physical act of constructing the table makes use of the connections between the hand and the brain; *kinesthetic* learners will derive special benefit from actually writing out the table. *Spatial* learners will actually visualize the table when recalling the material (e.g., during an exam).

Because the problem has twelve parts (reactions), it is well suited to a round-robin, take-your-turn approach to constructing the table. This provides a built-in device for engaging the members of the group with the material *and* with each other. Students with good *interpersonal* and *intrapersonal* skills will excel at making the external and internal connections that are inherent in the group approach to the table.

There are fundamental *linguistic* skills built into the problem as well. Special words stand for concepts, e.g., electrophiles, nucleophiles, mechanism. Each of these needs to be talked through in the group. The mechanism is a way of telling a story: "the electrophile does this and the nucleophile does that and this is what happens." The story can be told in words, in patterns, in pictures, in symbols (chemical equations) and in tables. The concept can be presented in a variety of formats that will be more or less accessible to different members of the group, depending on the nature of their abilities. Finally, the concept of the "mechanism" could be the starting point for building a concept map (see Chapter Five).

As you review the problems for your workshop sessions, look for ways in which different kinds of learners can find an entryway to the concepts involved. Watch for those moments when someone says "Oh, I get it!"—this tells you that you've found the In Door for that student's learning style. This is much fun.

Bibliography for Chapter Four

Chen, J. & Gardner, H. (1997). Alternative Assessment from a Multiple Intelligences Theoretical Perspective. In Flanagan, D., Genshaft, J., & Harrison, P., *Contemporary Intellectual Assessment: Theories, Tests, and Issues.* New York: Guilford.

Dunn, R., Dunn K., Price, G. (1985). *Learning Styles Inventory and Manual.* Lawrence, KS: Price Systems.

Gardner, H. (1999). *Intelligence Reframed: Multiple Intelligences for the 21st Century.* Basic Books: New York.

Gardner, H. (1993). *Frames of Mind: The Theory of Multiple Intelligences*; Basic Books: New York.

Krause, L. (1998). The Cognitive Profile Model of Learning Styles: Differences in Student Achievement in General Chemistry. *Journal of College Science Teaching*, *28, 1,* 57-61.

Light, R. (1992). *The Harvard Assessment Seminars, Second Report: Explorations with Students and Faculty about Teaching, Learning, and Student Life.* Cambridge: Harvard.

McWhorter, K. (1998). *College Reading and Study Skills,* 7th ed. Reading, MA: Addison Wesley.

Thorndike, E. (1920). The Reliability and Significance of Tests of Intelligence. *Journal of Educational Psychology, 11*, 284-287.

Thorndike, R. (1997). The Early History of Intelligence Testing. In Flanagan, D., Genshaft, J., & Harrison, P., *Contemporary Intellectual Assessment: Theories Tests, and Issues.* New York: Guilford.

Chapter Five: Basic Learning Principles

We're not computers

We live in a world dominated by computers, so it isn't surprising that computer imagery strongly influences the way we look at the world. In fact, we often think of *ourselves* like computers, especially when it comes to studying. As proof, let's spy on typical, well-motivated students as they get ready for an evening with the books. We're likely to watch them grab a soda, sit down and open their texts, and earnestly begin to read page one of the chapter. These students, given their good intentions, may work hard to focus on the new concepts and to understand the material being presented. They may take notes or underline in their text, and so on. They work hard until they've reached the end of the chapter or their energy.

The next afternoon, they proceed with the next chapter in the same way. And the next day, it's the same. These students *are* putting in hours of study time, and are quite sure they are keeping up with the material. However, when they finally return to these chapters in preparation for an exam, they may be shocked to find out that little information has stuck with them—not at all what they expected from their persistent labors. The same can happen with lecture notes; dutifully and accurately transcribed records of the lectures can look pretty unfamiliar after sitting in the notebook untouched for a few weeks, or even a few days. This happens because students expect their brains to function like biological computers. If we did have computers for brains, we could process information by carefully reading or taking notes and then pushing "Save." This would be great; come test time, we would just need to re-open our document-in-the-brain all the information would resurface, just like we entered it, ready to use. We would all look like geniuses.

But our brains don't work that way. Cognitive psychologists and neurologists know that certain requirements must be met most of the time if we are to hang onto information and make it significant and useful for us later. One of the requirements for reliable retrieval is that new information must have *meaning* for us. If you were asked to learn a long list of numbers, you probably could retain this list in your memory via brute strength—for a brief time. But unless you were able to invent some sort of personal connection to these numbers, you probably couldn't remember them for long, even though there is nothing much about digits you don't understand by now.

In order for new knowledge to be truly understood and retained, it also needs to be *tied to previous learning*. Without this, trying to learn is like holding an airplane model in one hand while you try to glue parts on with the other. Not only is this a clumsy way to do things, but you are likely to drop the whole thing as soon as someone tries to stack another set of pieces (i.e., more new knowledge) onto the one you are already holding. However, if you place the model on a steady table, this sturdy foundation will allow you to keep adding new parts for a good long time, as long as each piece is glued in firmly. We learn best by building on what we know.

And we learn well when our *attitudes* and *feelings* about new tasks are pitched at the right level. If we aren't engaged with the material at all, it's not likely that this information will stay with us over time. On the other hand, if we're too nervous about doing well, our anxieties can take on a life of their own and create an effective barrier to learning and retrieval. All of us have experienced the latter at some point, when we've been unable to call up a fact or an idea when we're under pressure. This is one of the reasons why 911 services have installed Caller ID on their systems; people in an anxious state of mind often can't retrieve their own addresses, let alone the complexities of stereochemistry or Coulomb's Law.

On the other hand, when we really enjoy something, we can learn vast amounts of information, almost without knowing it. For example, think of the intricacies of many on-line video games. In order to play, you need to know how to make internet connections to the game, you have to learn the rules of the game itself, usually at many different levels, the capabilities of different characters, etc. Much of this information is just about as complex as learning chemistry or math, but, for gamers, learning how to play is not only possible, it is half the fun. So *complexity* of information itself isn't always a barrier, but a lack of connection can be. Unfortunately, introductory science information can seem like a long list of useless facts to beginning students. Students don't have much context for this new learning, so it's easy for these facts to fall out of their heads almost as quickly as they are entered. This is where workshop leaders can play a truly important part; you can help your students see how you have used this information in your more advanced classes, in a lab, or maybe even in your own research projects.

Another important consideration is that new information be *put to use* fairly frequently, the "use it or lose it" phenomenon. For proof, think about all those facts you learned in your earlier years that are now lost to time. We'll bet you memorized the capitals of the states at some point, or the major battles of the Civil War, or the bones of the foot. Being the bright folks you are, you probably handled these learning tasks very well at the time. But, unless you have had reason to work with this material on occasion (like showing off at parties that you know Pierre is the capital of South Dakota), you've probably lost a considerable amount of this old information. However, if you have had reasons for repeatedly retrieving and using this material, it is probably integrated into your knowledge base. *Spaced practice and review* help you remember and internalize new information. It becomes a tool for you to use at will.

For some types of tasks, *social interaction* is another essential part of the learning process (Ryan & Stiller, 1991). Solo study with the books is a part of grasping new ideas, but often this is a necessary, but not sufficient, element of learning. In a study of successful students in science and math at selective institutions, a common factor among these "A" students was their ongoing connection to academic partners. They reported a close academic relationship with someone: an instructor, a TA, or fellow student (Light, 1992). It isn't hard to see why this might be the case; as we talk through concepts and problems with someone else, the gaps or misconceptions in our own learning become apparent, and, importantly, we get another pass through the material, thus satisfying the spaced practice criterion described above. Also, by explaining to others, you have put the material into your own words. When you teach something, you own it.

Workshop Strategies

So what can a leader do to ensure that these learning conditions are well established in the workshop? In answer, it should be noted that the very nature of the model itself speaks to these issues, i.e., workshops offer additional practice with the material, they appear to reduce anxiety and improve motivation (Black & Deci, 2000), and the sessions naturally provide an environment for focused social interaction about the material. However, a good leader also has a bag of specific tricks to help keep students' brains in action. Building ***concept maps*** is an effective way to highlight connections and relationships between ideas (see figure below). These maps are very easy to start: just write the main topic at the top or the center of a large sheet of paper or the blackboard. Distribute sticky notes and ask students to write whatever words come to mind and attach them where they seem to fit. Some fiddling and rearranging will be needed to get the connections among these ideas set up in a way that works, but that is part—maybe a major part—of the learning process.

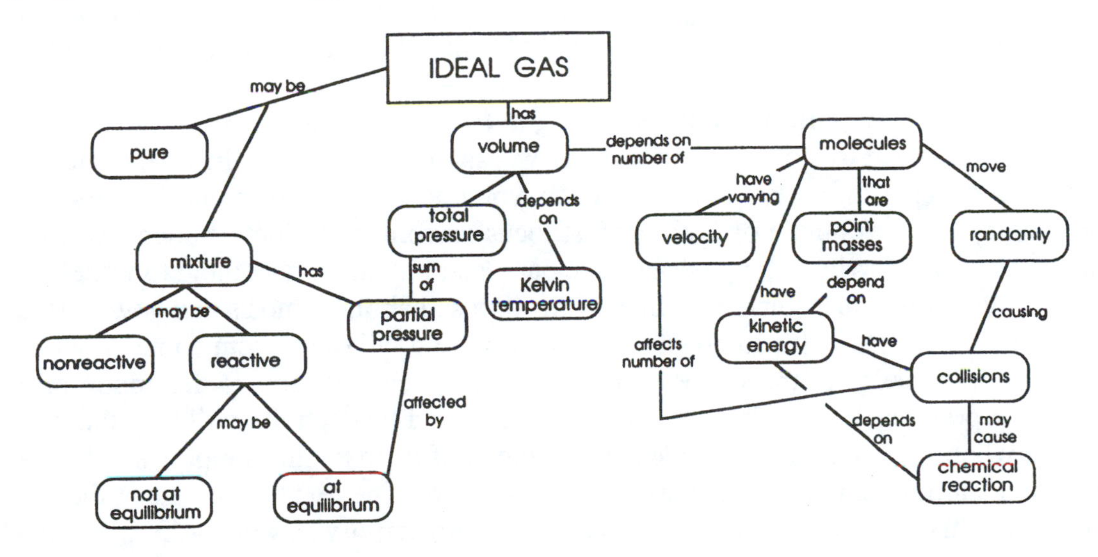

Figure 12.3. Concept map showing principles related to ideal gases.

from Herron (1996) p. 159

If you have used sketch paper, you and the students can take your maps home at the end of the session, and redraw them in more permanent formats. Distributing copies of the finished versions is a great way to review concepts at the next workshop session.

Model building is an important component of workshops, too. The application of this activity is most obvious in chemistry workshops, of course. Although we know that making models is essential to understanding molecular structure, we shouldn't assume that students will get the hang of these kits immediately. They may need help in using the kits—and they are almost certain to need instruction about their *purpose*, i.e., testing a hypothesis. We make a good guess, based on the data at hand, and then we build a model to see if our idea literally holds together. So this activity is much more than merely playing around with Tinkertoys. Model building can include the use of other manipulatives; something as simple as foam balls and pencils can be used to show the phases of the moon. We have asked students to stand up and move around to demonstrate difficult concepts. They themselves become the model, and this kind of activity helps to revive flagging energy in the middle of a tough workshop session.

Flowcharts are great ways to capture the details of a sequence or procedure, like mitosis and meiosis, or the steps needed to use a spectrophotometer. These charts make thinking visible. They can make series of events more distinct, they clarify where alternate routes can be taken, and they are especially great for your students who are visual and logical/mathematical learners. Flowcharts can be brought into the group discussion in several ways: you can prepare them ahead of time, as a sort of summary sheet, or you can direct the whole group to create one as part of the workshop session, or you can divide the group into subgroups of two or three students who are given the task of sketching out flowcharts as a startup activity for that day's new workshop problems.

Another easy startup activity is ***writing to learn***. Individually or in pairs, students are asked to respond to a prompt keyed to the workshop problems of the day, something like "For a couple of minutes, write down how you might determine osmotic pressure. Don't worry about grammar or spelling, just focus on content." Some students will tell you that they can't write a thing because they are lost, but encourage them to write anyway, even if it is just something like: "I'm confused about osmosis and how to figure out osmotic pressures." It's likely that the act of writing will help them come up with some prior knowledge of the subject at hand—we've already discussed the importance of connecting previous knowledge with new learning—and this activity will help them pinpoint where their understanding leaves off and confusion begins, helping set the stage for a good learning experience in the workshop session. The text they create at the beginning of the session provides an effective wrap-up activity as well. A few minutes before time's up, ask them to go back to these paragraphs and see how much more they know about the topic than they did a couple of hours ago. This helps to bolster the learning itself, and it can reinforce their motivation to show up to the workshop session the next time. If they can see their own improvement, they'll want to come back.

Pair problem solving (Herron, 1996; Whimbey & Lochhead, 1996) offers a variation to regular group discussion and analysis of problems. To use this method, divide your group into pairs and designate one person to serve as the problem solver and the other as the listener. The problem solver starts out by reading the problem and then continues to talk his/her way through the solution. The role of the listener is *not* to solve the problem, nor is it to trounce upon errors made by the problem solvers. The listener, who is giving full attention to both the problem solver and the problem, requests clarification ("Tell me more about that") or may note missteps ("You forgot to apply a rule here"). But mostly the listener is there as a sounding board, encouraging the solver to keep talking as he or she tries out approaches toward understanding the problem at hand. The listener then takes a turn as the solver on the next problem.

The advantage of this approach is in the elaboration, that is, the problem solver puts everything in his/her own words, under the guidance of the listener. This is a good example of the way in which a learning structure (i.e., having a solver and a listener) makes thinking visible. The activity ensures that ideas are fully thought through and tested at each step in the process. This method can go awry, however, if the listener starts "helping" too much (it can be hard to keep quiet!), or if the listener's commentary comes across more like surveillance than support. So you'll want to provide your students with guidance about what it means to listen and how to give constructive feedback.

Inquiry-based learning is more than an activity; it is an entire pedagogical approach with a rich literature and research base. Hassard (1992) describes it as follows: "Inquiry is a term used in science teaching that refers to a way of questioning, seeking knowledge of information, or finding out about phenomena" (p. 20).

Inquiry-based learning is essentially learning by discovery. Jerome Bruner, a long-standing proponent of discovery learning, observes that with this approach, "The pupil, in effect, becomes a party to the negotiatory process by which facts are created and interpreted. He becomes at once an agent of knowledge making as well as a recipient of knowledge transmission" (1986, p. 127. For an early discussion of this idea, see also Bruner, 1961).

To understand this process, it is important to experience and model it. One of the activities we have found successful is an inquiry project for leaders about the phases of the moon. This topic works well because there is considerable research available about the misconceptions people have about the moon's phases, and we do not need special equipment to observe the moon. Students need only a readiness to learn and to be willing to get past the obstacles that are present in urban environments where a view of the sky is limited.

When asked about the phases of the moon, most of us are convinced that we understand the interaction of the sun/earth/moon orbital motion and why we see lunar phases. But think about it. Can you project yourself into space and look at the sun/earth/moon system and determine at what time of the day or night we would see a rising first quarter or a setting full moon in New York, Santiago, Tokyo, or the city of Accra in Ghana?

Related to the field of inquiry-based learning is the study of conceptual change, which includes the study of *mis*conceptions. The video, "A Private Universe" (1987), can help workshop leaders explore their own understanding of conceptions and misconceptions. The video captures interviews with graduating Harvard seniors about their understanding of the phases of the moon and the cause of seasonal change on our planet. The video clearly shows that the perspective of these new college graduates is often not much different than that of ninth grade high school students and that both groups have misinterpreted their personal observations and their classroom learning in remarkably similar ways. Compare these two explanations of the earth's seasons, the first from a graduating Harvard senior—who had taken science courses—and the second from a ninth-grader:

Harvard senior:	*The earth goes around the sun and it gets hotter when we get closer to the sun and it gets colder when we get further away from the sun.*
Ninth grader:	*When it's winter is when the sun is furthest away from the earth and when it's summer is when the sun is closer to the earth.*

The video also shows how ineffective lecturing is in helping most students reach a clearer understanding and how important it is to help students actively construct their own learning. Using suggestions from the *A Private Universe Teacher's Guide* (1994), we have asked workshop leaders to write out and draw their own understanding about the phases of the moon, and then watch the tape. This activity helps leaders to capture their own existing mental models for these natural phenomena, what Posner and Strike (1992) would call the "conceptual context" in which their ideas have taken form. During the three-week unit, leaders work in research teams, record their observations of the moon, look into literature on the topic and share their ideas and their observations with their research teams. To understand the inquiry process and some of the research on misconceptions more thoroughly, we provide the workshop leaders with two articles: "Earth in Space" (Driver *et al.* 1994), about research into children's ideas regarding natural phenomena, and Duckworth's (1987) essay on "Learning with Breadth & Depth" from her book *The Having of Wonderful Ideas.*

After three weeks of observations, inquiry, and discourse within the teams, the groups present their current understanding of the sun/earth/moon system. The teams use diverse presentation tools to communicate with each other about their now-revised understanding of the interactions of the system, like concept maps, three-dimensional models, using "actors" to portray the systems through movement, or graphics on paper or computer. Typically, some of the presentations show an incomplete understanding, whereas others present sophisticated connections of the system. The progress often depends on prior knowledge and ideas, so we look back at leaders' original ideas and discuss the learning process they undertook.

The goal of this unit is to model for our leaders the conditions needed for conceptual change, i.e., what it takes to help people let go of fuzzy or wrong ideas and move towards sharper, more robust models of the ways the world works. Hewson (1996) suggests that, for real change and development of ideas, learners need to work within the following conditions:

1. A new concept needs to be *intelligible*. The only thing you can do with information you don't really understand is to memorize it by brute force. This kind of learning doesn't do much to build new ways of thinking; we need to understand before we can genuinely incorporate an idea into our mental models of the world.

2. It also needs to be *plausible*. A new idea needs to seem like a good fit for our internal database on the phenomena under review. Learners need the feeling that "Oh yeah, that makes sense" as they build new models of thought.

3. New concepts should appear *fruitful*. Sharper thinking helps us solve problems we didn't know how to address before; it also helps us ask questions we didn't think to ask. The value of a new concept to students has a great deal to do with how well they remember and use it.

We believe that the workshop offers a terrific format for promoting the intelligibility, the plausibility, and the value of new concepts for students. As Posner and Strike (1992) note: "A great deal of inquiry involves much talk. Explaining, arguing, constructing metaphors, giving counter examples, and the like express the social character of rationality" (p. 170). Sounds like a good workshop session!

Bibliography for Chapter Five

Black, A. & Deci, E. (2000). The Effects of Student Self-Regulation and Instructor Autonomy Support on Learning in a College-Level Natural Science Course: A Self-Determination Theory Perspective. *Science Education*, *84*, 6.

Bruner, J. (1986). *Actual Minds, Possible World.* Cambridge: Harvard.

Bruner, J. (1961). The Act of Discovery. *Harvard Educational Review, 31*, 1, 21-32.

Driver, R., Squires, A., Rushworth, P., & Wood-Robinson, V. (1994) *Making Sense of Secondary Science-Research into Children's Ideas*. London & New York: Routledge.

Duckworth, E. (1987). *The Having of Wonderful Ideas & Other Essays on Teaching & Learning*. New York: Teacher College Press.

Hassard, J. (1992). *Minds on Science*. New York: Harper Collins.

Herron, J. (1996). *The Chemistry Classroom: Formulas for Successful Teaching.* Washington, D.C.: American Chemical Society.

Hewson, P. (1996). Teaching for Conceptual Change, in *Improving Teaching and Learning in Science and Mathematics*, Treagust, D., Duit, R., & Fraser, B., Eds.

Light, R. (1992). *The Harvard Assessment Seminars, Second Report: Explorations with Students and Faculty about Teaching, Learning, and Student Life.* Cambridge: Harvard.

A Private Universe: Teacher's Guide (1994). Harvard-Smithsonian Center for Astrophysics, Science Education Department, Science Media Group, and the Corporation for Public Broadcasting.

A Private Universe (1987). Harvard-Smithsonian Center for Astrophysics, Science Education Department, Science Media Group.

Regis, A. & Albertazzi, P. (1996). Concept Maps in Chemistry Education. *Journal of Chemical Education*, *73*, 11, 1084-1088.

Ryan, R. & Stiller, J. (1991). The Social Contexts of Internalization: Parent and Teacher Influences on Autonomy, Motivation, and Learning. *Advances in Motivation and Achievement, 7*, 115-149.

Strike, K. & Posner, G. (1992). A Revisionist Theory of Conceptual Change, in *Philosophy of Science, Cognitive Psychology, and Educational Theory and Practice*, Duschl, R.,& Hamilton, R., Eds.

Whimbey, A., & Lochhead, J. (1986). *Problem Solving and Comprehension*, 4th ed. Hillsdale, NJ: Erlbaum.

Chapter Six: Learning Theory and the Workshop Leader

The world of learning theory and cognitive science can seem far afield from the lines of thought we commonly pursue in chemistry, biology, physics, and so on. Why should we look into learning theories at all? Why should a workshop leader care about educational psychology?

We care because the learning curve in introductory science and math courses is steep, and a semester is a short space of time. Since we are asking our students to master a great deal of complicated information under pressure, it makes sense that we consider what the experts in learning have to say. Within the rich field of cognitive psychology, we have been pleased that a number of respected cognitive theories connect well to the workshop model. The dynamics that happen during workshop sessions are complex, so no single theory explains all. (Sorry, we haven't located a Grand Unified Theory of Workshops.) What we do have, however, is a set of solid learning models that can explain a portion of what happens, or can fail to happen, in this model. We've summarized three theories that have special application to workshops, one in detail, the other two more briefly.

The Perry scheme and the workshop

An especially useful model for workshop leaders is the Perry scheme of college student development (1970). This set of ideas has provided a generative spark for several decades of further investigation.

Perry began his work in the 1950s at Harvard University, where he was the director of the Bureau of Study Counsel. Through a series of structured interviews with students during their four years of college work, he discovered a series of developmental steps through which many students appeared to progress.

Perry's team was able to replicate these findings, so they eventually proposed a model of nine developmental stages during the college years. The first three stages, which he groups under the label *The Modifying of Dualism*, describes a student's first encounters with multiple ways of looking at the world. These are students whose perspective, what you might call their "epistemological infrastructure," sets up clear boundaries between truth and error. They are not yet equipped to recognize the gray areas of knowledge and values. Instructors are seen to be in charge of the correct facts about their disciplines, and it is the students' responsibility to place identical versions of these facts in their own memories. Good instructors are those who lay out the Truth in the most straightforward way possible.

Perry quotes one student at this position: "A certain amount of theory is good but it should not be dominant in a course....I mean theory might be convenient for them, but it's nonetheless—the facts are what's *there*." (p. 67). In the workshop, students at this position may be overheard saying something like: "What is the point of doing these workshop problems if you never ever find out from the professor if you have the right answer or not? This seems like a pretty rotten thing to do to us!"

Students in this position can dismiss the value of authorities with little perceived experience. Perry notes that these students may regard a teaching assistant as "an outright fraud, a kind of older-brother pretender who arrogates the perquisites of Authority without its justification in knowledge" (p. 68). As a workshop leader, you may find yourself coping with a certain amount of disdain from such students, who believe that the professor (or maybe the textbook) is the only authority to whom credence can be given.

At this stage of intellectual development, students may be especially attracted to the sciences because these introductory courses seem to reflect a way of knowing that is precise, organized, and certain. So they may not appreciate any weighing of options on the part of the workshop leaders either. To them, a refusal to supply right answers to problems can look like the leader is dodging the issue or is not quite up to snuff with the material. This can be a little hard on your nerves—and you can find yourself slipping into the "teaching the class" position if you let this get to you.

Many students, Perry notes, enter college having progressed beyond this beginning stage. But even those who still hold a dualistic, right-wrong perspective on the world are soon confronted with multiple points of view, both in and out of the college classroom. They begin to perceive the nature of multiplicity, i.e, that for nearly everything in a college curriculum, there exist different points of view. While acknowledging multiple perspectives about some things, they may continue to believe that, while the real truths may be difficult to reach, there are definable certainties to be had, at least in some disciplines ("The Truth is Out There"). The nature of scientific inquiry encourages students to believe that scientists, through toil and dedication, are continually closing in on the Absolutes of the universe. Literature and art classes may appear to be soft and lacking in substance to these students; interpretation and the weighing of textual evidence do not yet make meaning for them.

Awareness of the diversity of human knowledge opens the door to new ways of knowing, so many students move from a dualistic point of view towards one in which relativism is seen as valid. Perry calls this middle group of positions the *Realizing of Relativism*. Some students at this point develop an "anything goes" attitude that can come across as pretty cynical. These students believe that, since Authorities do not appear to know everything after all, then, "Everyone has a right to his own opinion" (p. 97). In the workshop, our goal is to convey that, while we all may be entitled to an opinion, not all opinions are created equal!

Other students at this point may see independent thinking as a task set before them by the authorities in charge. Workshop leaders may find that students at this stage are willing to work through exercises designed to promote independent thinking, but they still expect True Answers to be revealed by the end of the session. And they can be quite peeved if you don't whip out an answer key so they can find out if they have arrived at the Truth. Resist the temptation to give in here. Instead, help them see that creating a *chain of reasoning* about an answer is the essential value of the workshop.

In the middle of Perry's scheme, the student experiences a transition period so crucial that he terms it a "drastic revolution." At this point, the student becomes aware that relativism is "the common characteristic of *all* thought, *all* knowing, all of man's relation to his world" (p. 111).

Workshop students may feel unsettled by changes among the fundamental relationships between themselves, authorities, and knowledge. Some, overwhelmed by these changes, may choose for a time to hold still and let this new understanding soak in. For others, this revolution is a big charge, so they find the energy and courage to develop a clearer understanding of the responsibility individuals have towards their own knowledge and understanding, even without authoritative rules about how to do so. This is the beginning of the third group of steps of Perry's model, the *Evolving of Commitments*.

Choice-making takes center stage with students at this point. Workshop leaders themselves may have gone through this change within the preceding year or two of their lives. They have progressed through some of the beginning stages of science instruction, and they have been college students for awhile, so they know that we all have to accept responsibility for our choices, whether the decisions are about the right answer to problem #4 or about choosing a major or a career. We can't expect the Authorities to make our lives for us.

Some workshop leaders may still find themselves in the midst of this decision making. If this describes you, leading a workshop can help you. In the immediate sense, you can see that students need to construct their own understanding of the course content. You could tell them each and every one of the right answers for the workshop problems, and this probably wouldn't help; in fact, it would probably impede their learning. You know that the real learning is about the *process* of coming up with answers and finding justification for these conclusions. When we feel satisfied because *we* know our answers make sense, we've learned something real.

Leading workshops can help you with decision making in the larger sense, as well. You find out if you flourish in a teaching environment, for example, or if you feel motivated to explore the cutting edge of the discipline you are helping your group understand. Keeping a log or a response journal to document the progress of your workshops can help you develop a more reflective stance towards your experiences.

Part of making decisions, large or small, is dealing with what happens next.

Coming to a conclusion is not the same thing as living with it. We all need what Perry refers to as "moral courage," in order make commitments in spite of a world in which knowledge is uncertain and our understanding is incomplete. As he notes, students may muster the "capacity to tolerate paradox in the midst of responsible action" (p. 166), and in fact, may discern that few of the resolutions we make in life are made once and then sealed for all time. Students can ultimately achieve awareness of the struggles we all face in reexamining and redefining the choices we have made. As one of Perry's students said:

> I sort of see this now as a natural thing—that you constantly have times of doubt and tension—a natural thing in existing and being open, trying to understand the world around you, the people around you (p. 165).

Perry acknowledges that some students do not progress in an orderly, predictable fashion from one stage to the next. His model characterizes three *Alternatives to Growth*. The first of these, *Temporizing*, is the least noxious in terms of students' future progress. This is a pause of a year or more, during which a student may be digesting a new understanding or waiting for the right motivation to make the next move.

A withdrawal to earlier stage of responsibility and commitment may have more serious implications. Perry calls this *Retreat*. Students who pull back into dualistic, right-wrong thinking have polarized their relationships with their professors and teaching assistants into an "Us against Them" position. Workshop leaders may recognize the student Perry quotes:

> Some of the teachers—I mean, it's more like a battle here between the students and the teacher to find out. I mean, see, you're here to learn, as far as... and I don't care how you get the information; one way or another, you're supposed to be able to learn, how to get information, and how to study it, and acquire it. But the teachers sure, sure don't, don't-ah, go to the trouble, much. They seem to try, to try, sort of go over things, go over it very briefly and let you find out the rest which is good, in a way, but you feel as if they're trying to hide a lot of facts (p.186-7).

Other students avoid the whole complex business of growth through *Escape*, or what Perry sees as the "defeat of care" (p. 200). These students are meandering through their lives. One of Perry's "escaped" students said, "As I get interested in something I probably *will* like it. Right now, I, I find it too easy to sort of goof around and not do anything. But I don't have any real interest in any of my courses... and, ah... and so it makes it sort of... I don't know" (p. 191). Workshop leaders may react with a sort of despair of their own in response to students who have chosen to escape the alternatives to growth. If your students aren't happy, it's easy to read this as evidence of inadequate leadership or as an indication of a student's lack of intelligence or moral fiber. Perry sees that the educator's responsibility is not to prevent students from escaping into alienation, but rather to maintain care and support while it occurs. Escape ultimately is not very satisfying, and so the despair and guilt these students feel may get them moving again. The opportunity to support students and provide a culture of growth is especially good in workshops, so leaders often witness students who get "unstuck" during the semester. Feeling like we've made a contribution to this development can be very gratifying.

As you might expect, the Perry scheme has been under review in recent years, given the limitations of the original sample from which it was developed, i.e., largely affluent white males, more than a generation ago. Educational theorists (e.g., Gilligan, 1982; Belenky et al., 1986; Gallos, 1995) have studied and then reworked the model to account for the experiences of other demographic groups.

One notion that has evolved from these recent studies is the concept of *Silence*, a stage even earlier than the dualistic, right vs. wrong beginning position of the Perry scheme. Belenky's team describes Silence as a time when self doubt overwhelms students' ability to think of themselves as real members of the group. Considering the cultural impact of race and gender, and the scarcity of women and minority science and math professors, it isn't difficult to see why it has often been hard for women and students of color to picture themselves as future professionals in these fields. These students can be shut down easily in traditional educational settings, feeling like they don't have a natural place at the table, so to speak.

Another point of view is that the aggressive "thrust and parry" of traditional academic debate can be off-putting for other groups of students. For students who operate from an "ethic of care" (Gilligan, 1982), a competitive approach to academic life can seem foreign and undesirable. Think about it from a family perspective. A family is probably not working very well if it is based on competitive dynamics, with some members emerging as winners and the others as losers. A family succeeds when everyone feels attended to and nurtured. So, to students who believe in an ethic of care, the traditional curricular format—where it often seems like the goal is to separate the academic victors from the washouts—can make science and math look like a rigged game.

The workshop model can respond beautifully to different groups of students. The natural activity in the sessions is for everyone to join in the academic debate, the weighing of options, and the demand for evidence and rationale to support opinion. But the environment promotes the goal of group care: understanding is achieved through the efforts of students working together for their common good. And, given the support and guidance offered by you as a leader, all students should feel that they have a legitimate voice in the proceedings; everyone should feel essential to the group.

Vygotsky and the workshop

Lev Semanovich Vygotsky (1896-1934) was a Russian psychologist who concluded that intellectual development is dependent on social interactions between student and teacher and among the student's peers (1987). Applying his theory, we see that peer leaders are the linchpins of the successes of the workshop model. You fill a role that cannot as effectively be filled by faculty or graduate teaching assistants, if we apply Vygotsky's theory to the workshop model. This is because students learn at the maximum level of efficiency when they work within what Vygotsky termed as their *zone of proximal development* (ZPD). This zone describes new concepts that the student is capable of learning. The lower end of the ZPD is defined by concepts that the student can learn without help; no real growth is involved. Beyond the higher end of the ZPD are tasks that the student cannot learn, even with assistance. Within the ZPD, our target teaching area, is the level of understanding for the students that is slightly above their present level; it is a stretch, but a realistic stretch. The result is growth. Workshop leaders are ideally suited for assisting student learning because they have a ZPD just slightly beyond that of their students. Who can better understand a student's ZPD than a peer who has recently learned the concepts him or herself?

Another key idea from Vygotsky is that, before students internally "own" new and complex ideas, they often are able to discuss these concepts at high levels. In other words, talking about higher-order concepts precedes true understanding of those concepts. This is where cooperative learning groups and challenging workbook problems come into play. The problems are designed to promote discussion among the students so that they can talk and listen to the ideas of others and then incorporate the conversation into their own understanding. You are a key to drawing students into this type of higher-order conversation and to ensuring the participation of each student.

Finally, Vygotsky also emphasized the importance of gradually removing support and allowing students more and more responsibility for their own learning as they become ready. Again, workshop leaders are ideally suited for this task, which Vygotsky called "scaffolding." At the beginning of the term, leaders provide a great deal of support, breaking problems into smaller steps, leading students toward correct solutions, and providing emotional support and encouragement for their students. As the term progresses, leaders gradually turn the responsibility for discussions over to their students. This process leads to the development of true life-long learners.

Deci and Ryan's motivation theory and the workshop

We all wonder why people do the things they do. What creates the pulls and tugs towards certain kinds of behaviors (like studying), and what causes us to avoid doing something?

Motivation can be seen as the engine that drives us--or fails to. When instructors and TAs are asked, "Why are some of your students doing poorly?" they often answer, "Because they don't seem motivated to study."

Trying to convince others to do something often promotes attempts at *external* motivation, the use of rewards and punishments to induce someone to take action. In the world of scholarship, external motivations include grades, both good and bad, praise from the instructor, parents, and friends, the fear of failure, the desire for a high salary, acceptance to graduate or professional school, and the like.

Over the long haul, many of these exterior inducements don't work very well, although they are certainly embedded in our culture. Carrots and sticks can get us moving for a time, but they will seldom propel us into a lifetime of deep involvement in an area of study. That requires *internal* motivation. For most of us to be happy, we need to hurl ourselves headlong into something, forgetting about whether this interest and commitment is going to reward us at a later date. You know you're on a roll with a particular subject or activity when you can't wait to get back to it. With real internal motivation, you forget to watch the clock or count the pages you have left to read. We're hoping you've arrived at this point in your own work; it's the key to long-term personal and professional satisfaction.

For some time, psychologists have been working to determine how interest and commitment toward particular activities are constructed. Of real help to the workshop leader is the concept of *self-determination* developed by Deci and Ryan (1991). It is their view that humans are born with the tendency to seek out opportunities to "master and integrate new experiences" (p. 230). They note that the social environment in which we live can either foster or impede the integration of the self. Fortunately, in their view, we can identify the factors that make specific social contexts helpful or unhelpful in this regard. One of these factors is the promotion of *competence*, or the sense that we are getting a grip on how to do things. A second factor is the support of *autonomy*, the feeling that we are our own agents or the origins of our behaviors. Yet another factor is the need for *relatedness*, which they define as "a person's strivings to relate to and care for others, to feel that those others are relating authentically to one's self, and to feel a satisfying and coherent involvement with the social world more generally" (p. 243).

These factors are successfully addressed by the workshop model. When this format is appropriately implemented, students gain a sense of expertise with a complex body of information (*competence*), they speak for themselves and their own interests in these sessions (*autonomy*), and they do so in an atmosphere that is socially supportive (*relatedness*). There is no question that the group leader is central to promoting the mastery, autonomy, and social connections that are integral to the development of intrinsic motivation.

Here are a few ideas about how you can promote the development of *internal* motivation to learn among your group members:

1. Help them see how far they have come in learning the subject at hand. Too often students are so intimidated about the material coming up in the academic term that they forget to look back to see how much progress they have made already. You can help

them do this from time to time, ensuring their sense of growing competence in the field.

2. Talk to them about their reasons for being in college and their future plans. Help them focus on their reasons for taking on this course and tackling this material. Students who are clear about their overall goals are ready to invest more heart in the workshop, making the sessions more effective for themselves and for everybody else, too.

3. Once they have solved a problem, assist them in articulating their thought processes to the others. This is good reinforcement for them, and good teaching for the others.

4. Compliment them on their supportiveness of each other. Social connections really matter in this model, so you'll want to make note of their efforts in this direction. For example, quiet, private remarks at the end of session like "I liked the way you helped Steve and Barbara with the second problem" can be very effective ways to promote group cohesion.

There is much more to be said about motivation models and many other educational theories, of course, but we hope that this introduction proves useful for you in the workshop. We encourage you to explore the discipline of cognitive psychology in more depth. We think you'll find this a rich source of information, as you alternate the hats you are currently wearing, that of a leader and that of a student.

Bibliography for Chapter Six

Belenky, M.; Clinchy, B.; Goldberger, N.; & Tarule, J. (1986). *Women's Ways of Knowing: The Development of Self, Voice, and Mind.* New York: Basic Books.

Deci, E. & Ryan, R. (1991). A Motivational Approach to the Self: Integration in Personality. In R. Dienstbier (Ed.), *Nebraska Symposium on Motivation: Vol. 38. Perspectives on Motivation.* Lincoln, NE: University of Nebraska.

Gallos, J. (1995). Gender and Silence: Implications of Women's Ways of Knowing. *College Teaching, 43*, 3, 101–105.

Gilligan, C. (1982). *In a Different Voice: Psychological Theory and Women's Development.* Cambridge, MA: Harvard.

Perry, W. (1970). *Forms of Intellectual and Ethical Development in the College Years.* New York: Holt, Rinehart, and Winston.

Vygotsky, L. (1987). *The Collected Works of L .S. Vygotsky.* Robert W. Rieber & Aaron S. Carton, Eds., translated by Norris Minick. New York: Plenum.

Chapter Seven: Race, Class, Gender, and the Workshop

Let's start this chapter out with two statements of belief:

#1 Everybody in your workshop deserves an equal chance to do well in the course.

#2 A small study group like a workshop can help equalize opportunities for all.

If academia were a simple and rational system, we could simply organize our workshops and make sure that all of our students attain a solid grounding in the content area and more confidence in their own abilities. The world isn't the least bit simple, however, especially not concerning issues like class, culture, ethnicity, and gender. Out in the open or not, these matters have much to do with how well a workshop leader is able to establish an environment that is comfortable and productive for all the students involved. Here's why:

1. *The demographics of science have traditionally been skewed.* We are only an eye blink away, historically speaking, from a time when the pursuit of scientific inquiry was almost *entirely* the provenance of those who could fit into the demographic path highlighted below:

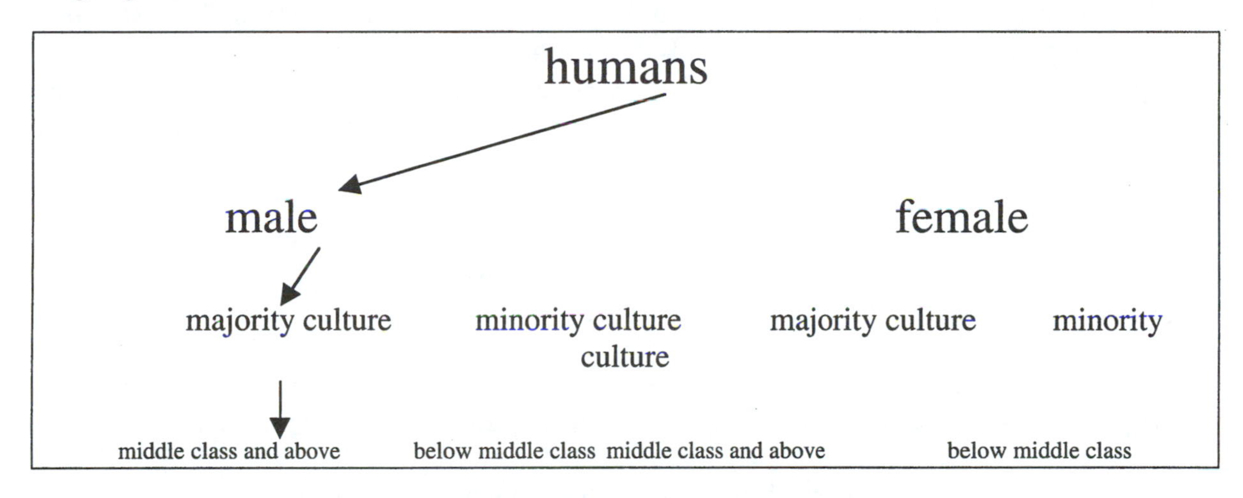

Everyone else? There was little need to apply.

2. *The current demographics of science still do not match the demographics of our society at large.* In case the diagram above seems like it describes a world long ago and far away, take a look at a recent headline from a recent National Science Foundation publications:

> **Despite Increases, Women and Minorities Still Underrepresented in Undergraduate and Graduate S & E Education**
>
> (National Science Foundation, 1999, p.1)

or this snippet—

> **Progress for underrepresented minorities in S & E graduate enrollment has been very modest. In 1975, they accounted for 3.7 percent of S & E graduate enrollment; by 1995, they accounted for 5.0 percent.**
> (National Science Board, 1998, p. 2-3)

3. *We don't check our attitudes at the door of the workshop room.* Social tensions cannot be blamed entirely on those who came before us. We have attitudes of our own about race, class, gender, and sexual orientation. And we don't check our feelings at the gates of our campuses; there is no guarantee that our colleges and universities serve as a neutral or "no fire" zone on these issues. We should assume that everyone brings beliefs about each other right into the workshop room along with books and notepaper and workshop problems.

4. Study groups have a great deal of potential for helping underrepresented groups in science, but conversely, *the close atmosphere of a workshop can also exacerbate tension.* In a lecture hall, everybody files in, sits down, and creates an audience, so to speak, for the upcoming lecture. The course format limits how much students are likely to step on each other's toes, culturally speaking. But in a workshop, there is plenty of such opportunity. People meeting together for a couple of hours weekly have many occasions during which they can launch little put-downs, ignore people, crack insensitive jokes, and the like.

So, what's a workshop leader to do? You can't be held responsible for sorting out the ills of the world, but you *are* charged with setting up an atmosphere in your group that creates a productive and welcoming workplace. To that end, here are a few suggestions:

1. *If you have a student who is the only one of his/her gender or culture in your group, pay particular attention to this situation.* Picture any of the following scenarios:

> A group is made up of all majority culture students, except for one African-American student.
>
> A single female student is in a group of male students.
>
> (Anticipating chapter 8) All of the members of the group are able bodied but one who uses a wheelchair.
>
> A single student from an economically poor background is in a group of students from more affluent homes.

It isn't difficult for students in these situations to feel shut down. In fact, the research tells us that isolated students are more at risk of dropping out of a group (Rosser, 1997;

Kemelgor, Neuschattz, Uzzi, & Alonzo, 1994). If such a grouping becomes evident at the beginning of the course, talk to your workshop program director.

2. *Pay attention to the little stuff, like jokes.* Although some outrageous and obvious comments and actions can pop up, most of the ways that people get to each other are quite subtle. A lot of it is in the form of humor—or what is supposed to pass for humor.

From a male student to a female student on her way up to the blackboard:

"Don't break a nail."

A story recounted by a workshop student:

On the week of the drag show on campus, I got to the workshop classroom early, and as I was coming in, I overheard these two guys talking about the show. They were laughing about it, I can't remember exactly what they were saying, but I know they said something like 'disgusting freaks' and 'fags.' I wanted to say something but I didn't have the guts... but now, whenever I see them I am very apprehensive, and I don't ever talk to them in workshop unless I absolutely have to.

Those on the receiving end of these remarks are stuck with a set of lousy options. They can laugh off the comments (and leave any bruised feelings unresolved), they can respond with reprimands (and come across as much too sensitive), or they can launch counterattack jokes and remarks.

All of these attacks and rebuttals, brief or extended, have a couple of things in common. First, they sidetrack the task at hand; the group is no longer analyzing the workshop problem before them, and the chemistry, biology, physics, or whatever is lost for the moment. More seriously, these little digs can be divisive. The jokers may think that they are loosening things up in a group, but just the opposite is likely to occur. If the person on the receiving end of the joke is embarrassed or offended, he or she is less likely to trust the others and work together. Before long, you have a group that is out of synch. As a leader, you may hope that ignoring comments and jokes will send the message that you disapprove. Actually, your silence may signal just the opposite: that you are in agreement with the speaker. Seymour & Hewitt (1997) quote a female student's story:

I was the only woman in a graduate-level physics class with seven men. They would tell jokes in bad taste, and watch to see how I handled it. Sometimes they would do really lewd things. They just did it to bother me. And, if I reacted, they would laugh at me until I would just want to kill them. But the professor would just ignore it. He wouldn't intervene to stop it, or to help me. He'd just say, 'Okay, let's get a move on,'—trying to make it as if nothing had happened (p. 245).

The bottom line here is that you need to take jokes and taunts seriously. If something like this happens in your group, you should deal with it publicly, with something like, "Please, no put-downs." If *you* are the one who slips up and says something potentially offensive, make your apologies right away. Common courtesy and common sense about how people feel can go a long way in these matters. (See Henes, McGuinness, and Clapp, 1994 for more ideas on this topic)

3. *Expect everyone to do well.* Without our conscious analysis, we can jump to conclusions about which students in our group are likely to succeed before they even open their mouths on the first day. These ill-founded conclusions have many negative effects. One consequence is on *wait time*, the term used for that time period after we ask a question and before we resubmit the question to another student or answer it ourselves.

Research in educational environments tells us that, if an instructor believes that student has a shot at doing well, he or she will provide more wait time for the student to puzzle through the answer. If we don't, this time is cut short (Stahl, 1994; Rowe, 1974). Consider the implications for students in the latter category. They may intuit at some level that such an instructor or group leader doesn't believe in their abilities to begin with, so question and answer sessions can already be dicey for them. Then, if these students have to pull answers out at twice the rate accorded to others, the game can seem hopeless. Pretty soon, the prophecy is fulfilled; without full engagement in the learning process, they perform less well, feel worse about it, talk less in group meetings—so the cycle keeps spiraling downward.

The answer to this is simple. Anticipate and act as though everyone in your group is a bright and capable person. It is great fun to watch students live up to this expectation.

4. *Remember that different conversational styles don't equal different abilities.* In academia, an assertive, confident, and direct communication style is usually the most rewarded. State your point, state your proof—this is what makes it in our educational world, and in the larger American culture. But not everyone approaches life that way. There are many cultures in the U.S. in which a bold approach comes across as rude or boorish. Students from these cultures may seem timid or incapable to those of the majority culture.

Similarly, much has been written about women's conversational styles (Tannen, 1991), and how they are interpreted. For instance, consider the tag question, a statement with a little question at the end:

We should use Gauss's law here, right?
There should be a double bond here, shouldn't there?
That's the Hardy-Weinberg Theorem, isn't it?

A negative interpretation of this style feature, a common element of women's speech, is that the speaker isn't sure of herself, and requires confirmation and assistance. To the woman, however, these tag questions may be her way of keeping the conversation going, by offering the listener an opportunity to pick up the next strand of the conversation. Far from being negative, this feature could be seen as a conversational component that helps glue a workshop session together.

5. *Help students who have been taught not to believe in themselves to change their minds.* By the time students get to college, they may have internalized many of the beliefs of the world around them, even if these beliefs are to their detriment. For example, women typically provide negative self evaluations about their math and science ability, even when exterior markers, like entrance exam scores and grades, tell them just the opposite (Seymour, 1997; Light, 1990).

This internalized negative view is probably the most entrapping of all the results of stereotyping and bias. What we say to ourselves is the most powerful of all forms of motivation. So you want to help by supplying students with better language. Take a look at these examples.

Student: *I'm just lousy in chemistry.*
Leader: *This material is kind of hard at first, but I know you can get the hang of it.* (Leader acknowledges what the student is saying, but indicates that the issue is with the intricacy of the material, not with the student's brainpower.)

Student: *I've never been any good at physics.*
Leader: *Boy, I really remember how stuck I was the first time I worked with B fields, too. Took me awhile to make sense of it. Here's what helped me get going.* (Leader uses him/herself as a role model and provides a strategy.)

6. *Accord credit fairly.* Your attention and approval are powerful tools. Consider the following true example:

A work group, headed by a female, completes a long-term assignment and submits it for approval. The reviewers ignore the woman who has done the lion's share of the job and turn to one of the men in the group and say, "Great job! We're delighted with how well you've completed this!"

We don't need a psychologist in the room to tell us how the woman feels, or how this affects her motivation to continue working with the project.

It seems trivial, but eye contact and pronoun use make a difference. If you have teams of two or three in your workshop working together on problems, look at *all* of them when you respond to their progress. This is one of the key values of a small group like a workshop; everyone can have the opportunity to build self confidence and skill in the subject.

In sum, our struggle in the workshops should be with the complexity of the concepts before us, not with each other. In fact, seen in the right light, the differences among us are truly assets to the group; they give us many points of view to focus on the problems at hand.

Bibliography for Chapter 7

Etzkowitz, H., Kemelgor, C., Neuschattz, M., Uzzi, B., & Alonzo, J. (1994). The Paradox of Critical Mass of Women in Science. *Science, 266*, 51-54.

Henes, R., McGuinness, N., & Clapp, L. (1994). *Creating Gender Equity in Your Classroom.* Davis, CA: College of Engineering, University of California, Davis.

Light, R. (1990). *Explorations with Students and Faculty about Teaching, Learning, and Student Life, First Report.* Cambridge, MA: Harvard.

National Science Board (1998). *Science and Engineering Indicators–1998.* Arlington, VA: National Science Foundation.

National Science Foundation (1999). Despite Increases, Women and Minorities Still Underrepresented in Undergraduate and Graduate S & E Education. *Data Brief.* January 15, 1999.

Rosser, S. (1997). *Re-Engineering Female Friendly Science.* New York: Teachers College.

Rowe, M. (1974). Wait-time and Rewards as Instructional Variables, Their Influence on Language, Logic, and Fate Control: Part One. *Journal of Research in Science Teaching, 11*, 2, 81-94.

Seymour, E. & Hewitt, N. (1997). *Talking about Leaving: Why Undergraduates Leave the Sciences.* Boulder: Westview.

Stahl, R. (1994). Using "Think-Time" and "Wait-Time" Skillfully in the Classroom. *Eric Digest*. Eric Document 370 885.

Tannen, D. (1991). *You Just don't Understand: Women and Men in Conversation.* New York: Ballantine.

Chapter Eight: Students with Disabilities and the Workshop

Think "disability," and we'll guess that an image of someone in a wheelchair comes to mind. Makes sense; the wheelchair icon is the symbol for disabilities used on doors, license plates, and signs in the parking lot.

Keep thinking, and you'll probably picture those with hearing and visual impairments, particularly if you've been in a classroom with a student who uses a sign language interpreter or a Brailler. But do you also think of those with invisible disabilities? These students number more than all of those with obvious disabilities put together. They might have a learning disability, attention deficit disorder, an emotional condition, seizure disorder, cardiac problems, and the like. (Just as a side note, remember that one disability doesn't vaccinate someone against having another. Some students, therefore, are dealing with multiple disabilities.)

Statistically speaking, there is a very good chance that you will encounter a student with a disability in your workshop. Laws and policies mandating educational access for those with disabilities are starting to do their job. For the official word on the legal side of things, see the Department of Justice Americans with Disabilities Act webpage listed in the bibliography. As part of a natural progression, these students are entering college in record numbers; enrollment of students with disabilities has increased three-fold since the late 1970s (Heath, 1995).

You may feel a little nervous at the thought of being responsible for making your workshop accessible to someone with a disability, but it is important to remember that these students have already achieved academic success; if they hadn't, they wouldn't be in college. Very often, they can tell you what aids and learning strategies work for them. And since the law requires that appropriate academic adjustments be made, there must be a coordinator of disability support on your campus; this person also can help you should the need arise.

All this said, it is still a good idea for you to do a little background work yourself, so here is a brief look at disability issues in the workshop. (See also the AHEAD web address at the end of this chapter.)

Mobility Support

Students who use wheelchairs will need accessible space and perhaps a table that is the correct height. If their upper extremities are impaired, they may need adaptive technology as well. These issues should be addressed by the appropriate college office before the workshop begins. If they haven't been, speak up—or encourage the student to do so. Most of the time, students who use wheelchairs will be able to use the blackboard more or less like anyone else. But if there is something about your room that makes this more or less like anyone else. But if there is something about your room that makes this difficult, take action. Get the room rearranged, or find some alternate way for the student to display his or her work to the rest of the group.

Visual Support

Students with low vision might require a computer program that enlarges printed material or equipment such as a reading machine that scans text and converts it to speech. The technology helps, of course, but it is still a good idea for you and the other students to talk through the charts, diagrams, and visual models as you use them in your workshop sessions.

Students with no usable vision will need materials in audio format or Braille and may need initial assistance in finding their way around the room. If a guide dog accompanies the student, please talk to the student, not the dog. Guide dogs always seem to fascinate others, but don't let the other workshop students tempt the dog into a game of fetch—this dog is at work. In fact, take care that the dog is not distracted in any way.

Be clear and concise when you are giving direction or explaining concepts because this student will be relying on memory. It might be helpful for the student to tape record the workshop sessions. Don't worry about the disruption of a tape recorder; after the first five minutes or so, everyone seems to forget that it is running.

Hearing Support

Students with limited or no hearing use a whole range of accommodations to help them build an interface to the hearing world. They may have a notetaker because they cannot simultaneously watch the interpreter or speechread and take notes. They might use adaptive technology like hearing aids, FM loops, some form of real-time captioning, and other newly-emerging technology.

It's important to remember that, most of the time, a hearing aid doesn't compensate fully for a hearing loss. Many hearing people figure that a hearing aid sharpens sounds, like using the tuner on a radio when the station isn't coming in very clearly. A better analogy is a comparison to the volume control on your radio. Despite recent improvements, wearing a hearing aid is like turning the volume up on a station that isn't coming in well; you get more signal, but you can get a lot of additional static, too.

Some students use sign language interpreters who are trained to convert speech to a visual format through sign language. Others may use an oral interpreter, who is trained to mouth the words of speakers, using an inaudible whisper. The student then speechreads the interpreter's lips. If you have an interpreter assigned to your workshop, you'll need to make certain that the room is arranged so that the interpreter can sit in front of the room facing the student who is deaf. The interpreter's *only* job is to enable the student to understand what is being said in the workshop, i.e., he or she is not an intermediary, a tutor, or anything else in this context. Although this may feel strange at first, conversation should be directed to the student, not the interpreter. The interpreter will also expect all the workshop participants to take turns in speaking and not talk over each other; he or she can't interpret more than one speaker at a time. All participants must speak clearly and distinctly so that the interpreter can do his or her job. As the leader, you may need to remind students of these requirements from time to time.

Support for Learning Disabilities

In terms of numbers, this is the largest college population covered by the Americans with Disabilities Act (Heath, 1995). Students with learning disabilities may have difficulty in understanding spoken or written language or in listening, thinking, speaking, or completing math calculations, despite otherwise normal cognitive ability and adequate previous opportunity to learn. Other students have learning disabilities that make it difficult for them to interpret other people's facial expressions and tone of voice, which can cause them to misinterpret the thoughts and feelings of their fellow workshop members.

One way to think about this is to consider the differences in learning styles we discussed in Chapter 4. If you were to prepare a profile for everyone in your workshop, you would see patterns of strengths and weaknesses for all of your students, but, for most, the pluses and minuses would fall within a moderately-sized bandwidth. For a student with a learning disability, however, the bandwidth can be huge. A learning disabled profile often has distinct peaks and valleys in test scores. In fact, this is how a learning disability is typically detected.

So the goal for the students—and by extension, for you as the leader—is to organize information into formats that correspond to strengths and to find compensations for the weak spots. Students may already have notetakers, recorded books, and computer-assisted technology in place. They may also benefit from sketching out problems and explanations during the session, the creation of concept maps and other ways to display information graphically, or more tactile methods, like building models and other hands-on activities. Obviously, these learning strategies can be great for *all* the students in your room. This is something that we've discovered about disability support in general: The work done to create access for someone with a special need often unexpectedly helps others as well, much like curb cuts for those in wheelchairs makes things easier for those pushing baby strollers. (Note: Much useful information about learning disabilities can be found in Lerner, 1997, a widely available text.)

Support for Attention Deficit Disorder

Not so long ago, ADD was thought to be a disorder of childhood; kids with this diagnosis were expected to outgrow it by age 12 or 15. We now know that this isn't true; ADD may look different for a twenty-year-old compared to a five-year-old, but the effects of this disorder are often felt across the lifespan.

This diagnosis refers to students who have difficulty changing focus as needed or maintaining their attention for any length of time. They may gaze out of window or be engaged in their own thoughts. They may become so engrossed in their homework that they forget to get up and go to class. In the workshop, these students may have trouble sitting still, they may talk out of turn, or be disorganized.

It can help to structure the workshop's activities and announce your plan of action at the beginning of the session ("Here's how I think we should go about things today"). Students with ADD may also need workshop space that is as distraction free as possible, so if your workshop is scheduled in a noisy room, talk to your program directors about the possibility of finding another location. Successful students with ADD have learned strategies that enable them to function well in academia, so let them guide you in setting up a good workshop environment and format.

Support for Emotional Disabilities

Like the population at large, college students may experience depression, anxiety, obsessive-compulsive thoughts, bipolar disorder, and so on. Given the success of treatment programs now available, often these emotional disorders (ED) can be cured or controlled, so many of those so diagnosed can do well in a college program.

How might emotional disabilities have an impact on your workshop? Despite their portrayals in movies and on TV, students with ED aren't likely to cause mayhem in the workshops, or anywhere else. To clarify, these students are not more likely to behave in a violent way than the typical student. In everyday life, this sort of disability actually makes someone more likely to be a *victim* of violence than a perpetrator (Unger, 1998).

But they may have short-term memory problems, causing them to forget directions, previously taught material, or workshop expectations. They may experience periods of self-absorption, and thus tune out the workshop. Participants may be distractible and have trouble concentrating on the task at hand. Or they may find that new environments and social situations are hard to deal with. On the physical side of things, some psychotropic medications are a challenge for the user, since they can cause fatigue, muscle aches, and excessive thirst.

Here are some strategies you can implement to help students with ED find success in your workshop:

1. Assist students with time management and study strategies that relate to this course. For example, you may want to show them how you tackled the chapters in the text (e.g., by taking notes, writing in the margins, making flashcards, color coding new words).

2. Provide breaks when appropriate.

3. Include a variety of methods for the presentation and explanation of problems.

4. Repeat instructions and explanations as needed—or better yet, get other students to go through things again.

5. Create summary sheets or overheads about key concepts.

6. Keep an eye on group dynamics. If the student with ED starts to become a loner in the group, think about ways to bring him or her back into the conversation.

When leaders read the list above, their typical reaction is something like, "Well, that's just good leadership in general." Exactly. Good disability support most of the time is just putting common sense—and a little technique—into practice.

A Few Other Notes

Remember that information about a disability is much like any other medical record, so think about confidentiality. The student with a disability is free to discuss these issues liberally, but, as a leader, you need to assume that this is private information. You may seek out assistance from the program directors or from your office of disability support, but don't discuss a disability in the group (unless the student has clearly initiated such a discussion), and you must never chat about an individual's disability in the dorm or in the halls, etc.

Please know that a workshop leader's confidence in all students is essential to their success. In many ways, the workshop format is one of the best forms of disability support we have available. The small size, the interaction, the opportunity to rehearse ideas with others, the use of different learning modalities, the social support—all of these things help students in general, but they may be indispensable for the student with a disability. It is a pleasure to make a crucial difference in someone else becoming successful. So if you have a student with a disability in your group, we are sure that the student will learn a great deal—and so will you. Enjoy yourselves.

Bibliography for Chapter 8

AHEAD, Association on Higher Education and Disability. http:///www.ahead.org

Heath Resource Center (1995). *College Freshmen with Disabilities: A Triennial Statistical Profile.* Washington, DC.: American Council on Education.

Lerner, J. (1997). *Learning Disabilities: Theories, Diagnosis, and Teaching Strategies*, 7th ed. Boston: Houghton Mifflin.

Unger, K. (1998). *Handbook on Supported Education: Providing Services for Students with Psychiatric Disabilities.* Baltimore: Brooks.

U.S. Department of Justice Americans with Disabilities Act Homepage http://www.dinf.org/crt/ada/adahom1.htm

An Anthology of Readings

The Workshop Chemistry Project: Peer-Led Team Learning*

David K. Gosser, Jr.
Chemistry Department, The City College of the City University of New York,
138th Street and Convent Ave., New York, NY 10031

Vicki Roth
Learning Assistance Services, University of Rochester, Rochester, NY 14618

My first day! At first, the butterflies in my stomach were all I could concentrate on. So I took attendance and said a few words about myself and organic chemistry so that the students would see me as part of the group and not some unapproachable Orgo genius.

I definitely have to say that my first day as a workshop leader taught me a lot about being an effective leader. There is more preparation than I had anticipated, and more patience required than I had thought. How could I get them to see that I was not there to dictate answers, but to act as their peer mentor?

I was shocked to see my most shy student taking a very active role.

I'd like to note my feelings as a workshop leader. The students encounter their first test Friday, and I'm nervous for them. Is this normal?

Since most of my students did not do well on the first exam, they were afraid of the second exam. I told them what happened when I took general chemistry a year ago. In my group, I got the lowest grade. Encouraged by my workshop leader, who told me that if I studied hard I could be a leader too ... I started to study harder, so I did well in the course and now I am trying to help other students as I was helped.

These comments, and many others like them, appear in the reflective journals kept by our peer leaders as they trace their experiences guiding students in Workshop Chemistry. A coalition of faculty, students, and learning specialists, the Workshop Chemistry project is developing a peer-led team-learning model for teaching and learning chemistry (1). The Workshop Chemistry model embraces dimensions of student experience that are essential for learning: the freedom to discuss and debate chemistry in a challenging but supportive environment, the connection to mentors, and the power of working as part of a team. The workshop model calls for the traditional recitation, or a modest amount of lecture, to be replaced by a *new curricular structure: a two-hour student-led workshop*. In the first two and a half years of the project, more than 6000 students have participated in workshop courses in allied health, general, and organic chemistry, conducted by 27 faculty and more than 800 workshop leaders.

The Workshop Model

In the workshop model, the class is divided into groups of six to eight students who work together throughout the term under the guidance of an undergraduate peer leader (1). The peer leader, called the Workshop leader, is a student with good

* *J. Chem. Educ.* **1998**, *75*, 185-187.

communication and people skills who has done well in the course previously. The leaders receive a modest stipend in recognition of their participation. In the Workshop model, the faculty reallocate some of their time and energy from presenting information to shaping a peer-led learning environment in which students are actively engaged with the subject. The Workshop model builds on earlier work and shares elements of group or team learning with other efforts (3-5) but has the unique characteristic of peer leadership as an integral part of the course structure.

Role of the Workshop Leader

The workshop leader is there to actively engage students with the materials and with each other. This facilitation can take many forms: organizing "round-robin" style problem solving, creating subgroups or paired problem-solving groups whose members can compare results, offering timely assistance when a group is stuck, and providing encouragement and guidance in the study of chemistry. The workshop leader needs to set a tone for the discussion in which individual points of view are respected, the criticism is constructive, and all members have an equal opportunity to participate. As the term progresses and as the leaders guide their students through the trials and tribulations of a difficult course, they often become mentors and role models for the members of the group.

Recruiting Workshop Leaders

Although it requires an investment of time, it has turned out to be a pleasure to recruit and train new leaders because it gives us a mechanism to provide new opportunities for our best students. We find most of our leaders for the next semester by inviting our top students to an interest meeting. During this session, we and our experienced leaders describe the project as a whole and outline the benefits and responsibilities of being a workshop leader. Although most of us were concerned at the outset that we would not locate a sufficient number of leaders to sustain the project, we have not found this to be the case. Instead, we generally find a previously untapped enthusiasm for taking on this kind of role and responsibility.

Training Workshop Leaders

Each semester our new leaders are energetic, smart, and eager to begin meeting with their own group of students. However, despite their natural talents and enthusiasm, it is important that we guide the development of their leadership skills. The natural inclination for many new leaders is to fall back into familiar patterns of instruction; we need to help them understand that their role is not to serve as an additional lecturer for the course or as a group tutor. Because the Workshop Chemistry consortium includes a wide range of colleges and universities, we have worked toward a flexible leader training format that ranges from a series of staff meetings between chemistry faculty and their leaders to credit-bearing training courses, team taught by chemistry faculty and learning specialists.

Within these varied formats, the training highlights the following:

Faculty and leaders review the content and practice of the workshop problems

Instruction and practice in group dynamics and group leadership

Discussions about learning styles and intellectual development

Practice with new pedagogical methods and review of study tactics

Discussions about the impact of race, class, and gender on learning environments.

In the training sessions, new leaders have the opportunity to receive feedback from us and from their peers, to pursue special investigations of topics that interest them, and to explore how their workshop experiences may lead them to future academic careers.

Materials for the Workshop

Good workshop problems can reinforce the goal of promoting collaborative, active engagement with chemistry. The structure of the problem, or the manner in which it is phrased, can assist the group interaction. The faculty in the project have developed workshop problems on a topical basis to form a collection for other faculty to choose from or add to. Feedback and suggestions from the leaders about the problems under actual workshop conditions have been very useful. We have used a variety of approaches in designing workshop problems:

Stepwise or structured problems, reflective problem solving

Construction of concept maps, simulations using concrete models

Interpretation of graphs, observation/deduction problems

Problems involving the use of molecular models

Workshops based on a laboratory experience

Problems based on historical developments in chemical thought

Problems related to important contemporary issues

Creating strategies for synthesis, moving from data to structure and mechanism

We do not write answer keys for workshop problems. The existence of an answer key would undermine the philosophy of the workshop, which is to learn how to construct answers.

Evaluation of the Workshop Chemistry Project

We have collected data from workshop courses at City College of New York, New York City Technical College, Borough of Manhattan Community College, Lehmen College, Queens College (New York), Medgar Evers College, and the Universities of Pittsburgh, Pennsylvania, Rochester, St. Xavier (Chicago) and the Pacific. The evaluation to date demonstrates that the Workshop model and the training component are robust (6). We have tracked Workshop Chemistry students across a number of different sites, and have found statistically significant improvement in grades, retention, and levels of student satisfaction. We have administered questionnaires and held focus groups with workshop participants, their leaders, and the professors in charge of these courses. All three groups have enthusiastically endorsed the workshop model.

The present focus of the evaluation is to determine the critical components of a successful workshop course. We have identified the need for leader training, coherent and challenging workshops, close involvement of the faculty with the leaders and the integration of the workshop with other course components.

We have also conducted a preliminary test (7) of the theory of autonomy support (8,9). In this theory, a learning environment that supports students' autonomy will promote their growth as learners. Our assessment work to date has borne this out; the degree of autonomy support reported by workshop students is positively correlated with their performance at the end of the term. It is this particular characteristic of peer-led support that is the heart of the Workshop model.

The View From Industry

We want to ensure that the workshop model serves the needs of students as they enter the workplace. Our own survey of industry and other studies (9-11) indicate that communication skills and teamwork are prerequisites for success in the workplace. The performance requirements that companies face demand a quality that can often be achieved only through collaboration and teamwork (12). From an industry perspective, a *high-performance team* has been defined (13) as a *small number of people who are committed to a common goal, working approach, and to one another's personal growth and success*. The successful workshop experience shares these qualities of collaborative work, personal connection, growth, and high performance.

Resources

The peer-led Workshop is a flexible and robust model for teaching and learning chemistry. It has been tested and validated throughout the Workshop Chemistry Consortium, at a wide variety of institutions. We invited others to adopt, adapt, and participate in the creation of materials and methods for Workshop Chemistry. The following materials are available through the Workshop Chemistry Project:

Topical workshop units in allied health, general, and organic chemistry

A guidebook for implementations of workshops

Materials for the training of workshop leaders

To help others develop workshops in their own courses, we offer interactive seminars at chemical education meetings on a regular basis, where our undergraduate leaders participate, leading workshops to directly illustrate the model to interested faculty and their students (14-19). In addition to such seminars in the future (20), we plan to host a series of more extensive four-day retreats in the summers of 1988 and 1999. To keep in touch with these and other developments, consult the Workshop Chemistry Home page at *http://www.sci.ccny.cuny./edu/-chemwkshp.*

Acknowledgments

We would like to acknowledge the students, faculty, and learning specialists across our coalition for their participation in the development of the workshop model. We are particularly grateful to Jack Kampmeier, Pratibha Varma-Nelson, and Victor Strozak for their contributions to the shared vision of the peer-led model. We thank Leo Gafney and A. Black for their work on evaluation. DKG gratefully acknowledges the work of the team partners in the development of the workshop at City College, Michael Weiner, Stanley Radel, and Ellen Goldstein. We are pleased to acknowledge support from the National Science Foundation, Division of Undergraduate Education, Course and Curriculum Development Program (NSF grants DUE 9450627 and DUE 9455920).

Literature Cited

1. Gosser, D.; Roth, V.; Gafney, L.; Kampmeier, J.; Strozak, V.; Varma-Nelson, P.; Radel, S.; Weiner, M. *Chem. Ed.* **1996**, *1*(1).

2. Woodward, A.; Gosser, D.; Weiner, J. *J. Chem. Educ.* **1993**, *70*, 651.

3. Treisman, U. Z. *Coll. Math. J.* **1995**, *23*, 362.

4. Cracolice, M.; Roth, S. *Chem. Ed.* **1996**, *1*(1).

5. Dinan, F. J.; Frydychowski, V. A. *J. Chem. Educ.* **1995**, *72*, 431.

6. Gafney, L. Project Evaluator's Report to the National Visiting Committee, **1997** (available upon request).

7. Black, A.; Deci, E. The effects of student self-regulation and instructor autonomy support on learning in a college-level natural science course: A self-determination theory perspective, *Am. Ed. Res. J.*, submitted for publication.

8. Deci, E. L.; Ryan, R. L. *Intrinsic Motivation and Self-Determination in Human Behavior*; Plenum: New York, **1985**.

9. Deci, E. L.; Eghrari, H.; Patrick, B. C.; Leone, D. R. *J. Personality* **1994**, *62*, 119.

10. Maxfield, M. In the ACS Seminar Manual, *Undergraduate Chemistry Curriculum Reform, Its Effect on High School and College Level Teaching*; American Chemical Society: Washington, DC, **1996**, p 87.

11. American Chemical Society *Committee on Professional Training Newsletter* **1995**, *8* (Fall), 1.

12. The Secretary's Commission on Achieving Necessary Skills (SCANS), *What Work Requires of School: A SCANS Report for America 2000*; U.S. Department of Labor: Washington, DC, **1991**.

13. Reich, R. B. *The Work of Nations*; Vintage: New York, **1992**.

14. Katzenbach, J. R.; Smith, D. K. *The Wisdom of Teams*; HarperBusiness: New York, **1994**.

15. Gosser, D.; Varma-Nelson, P.; Daniel, J.; Patitucci, D.; Munch, C.; Johanek, A.; Boeschel, C.; Semlow, S. Workshop Chemistry: Peer-Led Team Learning. Presented at Day 2 to 40 Workshop Symposium on Chemical Education, Ann Arbor, MI, **1997**.

16. Varma-Nelson, P.; Kampmeier, J.; Zhang, R.; Johanek, A. The Organic Chemistry Workshop. Presented at the Middle Atlantic Discovery Chemistry Project, Frederick, MD, **1997**.

17. Varma-Nelson, P.; Strozak, V.; Patitucci, D.; Boyle, M.; Boeschel, C.; Semlow S.; Pellegrini, S. A Peer-Led Team-Learning Model for Teaching Chemistry. Presented at Revitalizing General Chemistry at the Community College, PKAL, Oarkton, IL, **1997**.

18. Roth, V.; Varma-Nelson, P.; Kampmeier, J.; Strozak, V.; Gosser, D.; Zhang, R.; Daniel, J.; Boeschel, C.; Glenn, K.; Ambe, R.; Spolitino, F. Peer-Led Team Learning. Presented at the Improving the Teaching/Learning Process in General Chemistry Conference: Stony Brook, NY, **1997**.

19. Gosser, D.; Fernandez, L.; Sohel, M.; Liu, R.; Khadim, H.; Jacob, S. Transforming the Learning Environment for Science and Mathematics at the Urban and Commuter Institution. Presented at Workshop Chemistry, PKAL, New York, **1997**.

20. The Workshop Chemistry Project. A Chautauqua short course; Philadelphia, July 11-13, **1998**.

25 Ways to Get the Most Out of Now[*]

D. B. Ellis

The following time management techniques are about when to study, where to study, ways to handle the rest of the world, and things you can ask yourself when you get stuck. As you read, underline, circle, or otherwise note the suggestions you think you can use.

When To Study

1. Study difficult (or "boring") subjects first. If your chemistry problems put you to sleep, get to them first, while you are fresh. We tend to study what we like first, yet the courses we find most difficult often require the most creative energy. Save the subjects you enjoy for later. If you find yourself avoiding a particular subject, get up an hour early to study it before breakfast. With that chore out of the way, the rest of the day can be a breeze.

Continually avoiding a subject might indicate a trouble area. Further action is called for. Clarify your feelings about the course by writing about those feelings in a journal, talking to an instructor, or asking for help from a friend or counselor. Consistently avoiding study tasks can also be a signal to re-examine your major or course program.

2. Be aware of your best time of day. Many people learn best in daylight hours. If this is true for you, schedule study time for your most difficult subjects when the sun is up.

Unless you grew up on a farm, the idea of being conscious at 4 a.m. might seem ridiculous. Yet many successful business people begin the day at 4 a.m. or earlier. Athletes and yogis use this too. Some writers complete their best work before 9 a.m.

Some people experience the same benefits by staying up late. They flourish after midnight.

If you aren't convinced, then experiment. When you're in a time crunch, get up early or stay up late. The new benefits you discover might even include seeing a sunrise.

3. Using waiting time. Five minutes waiting for a bus, 20 minutes waiting for the dentist, 10 minutes between classes—waiting time adds up fast. Have short study tasks ready to do during these times. For example, carry 3 x 5 cards with facts, formulas, or definitions and pull them out anywhere.

A tape recorder can help you use commuting time to your advantage. Make a cassette tape of yourself reading your notes. Then play these tapes in a car stereo as you drive, or listen through your earphones as you ride on the bus or exercise.

---

[*] Ellis, D. B., *Becoming a Master Student*, 8th Ed., Copyright 1997 by Houghton Mifflin Company. Used with permission.

Where to Study

4. Use a regular study area. Your body and your mind know where you are. When you use the same place to study, day after day, they become trained. When you arrive at that particular place, you can focus your attention more quickly.

5. Study where you'll be alert. In bed, your body gets a signal. For most students, it's more likely to be "Time to sleep" than "Time to study!" For that reason, don't sleep where you study. Just as you train your body to be alert at your desk, you also train it to slow down near your bed.

Easy chairs and sofas are also dangerous places to study. Learning requires energy. Give your body a message that energy is needed. Put yourself into a situation that supports that message.

6. Use a library. Libraries are designed for learning. The lighting is perfect. The noise level is low. Materials are available. Entering a library is a signal to quiet the mind and get to work. Most people can get more done in a shorter time at the library. Experiment for yourself.

Ways to Handle the Rest of the World

7. Pay attention to your attention. Breaks in concentration are often caused by internal interruptions. Your own thoughts jump in to tell you another story about the world. When that happens, notice the thoughts and let them go.

Perhaps the thought of getting something else done is distracting you. One option is to handle that task now and study later. Or write yourself a note about it, or schedule a specific time to do it.

8. Agree with living mates about study time. This includes roommates, parents, spouses, and kids. Make the rules clear, and be sure to follow them yourself. Explicit agreements—even written contracts—work well. One student always wears a colorful hat when she wants to study. When her husband and children see the hat, they respect her wish to be left alone.

9. Get off the phone. The telephone is the ultimate interrupter. People who wouldn't think of distracting you might call at the worst times because they can't see that you are studying. You don't have to be a telephone victim. If a simple "I can't talk, I am studying" doesn't work, use dead silence. It's a conversation killer. Or short-circuit the whole problem: unplug the phone. Get an answering machine or study at the library.

10. Learn to say no. This is a timesaver and valuable life skill for everyone. Many people feel it is rude to refuse a request. But saying no can be done effectively and courteously. Others want you to succeed as a student. When you tell them that you can't do what they ask because you are busy educating yourself, most people will understand.

11. Hang a "do not disturb" sign on your door. Many hotels will give you one free, just for the advertising. Or you can make a creative one. They work. Using signs can relieve you of making a decision about cutting off each interruption—a timesaver in itself.

12. Get ready the night before. Completing a few simple tasks just before you go to bed can help you get in gear faster the next day. If you need to make some phone calls first thing in the morning, look up those numbers, write them on 3 x 5 cards, and set them near the phone. If you are set to drive to a new location, make note of the address and put it next to your car keys. If you plan to spend the afternoon writing a paper, get your materials together: dictionary, notes, outline, paper, and pencil (or disks and computer). Pack your lunch or gas up the car. Organize the diaper bag, briefcase, or backpack.

13. Call ahead. Often we think of talking on the telephone as a prime time-waster. Used wisely, the telephone can actually help you manage time. Before you go shopping, call the store to see if it carries the items you're looking for. If you're driving, call for directions to your destination. A few seconds on the phone can save hours in wasted trips and wrong turns.

14. Avoid noise distractions. To promote concentration, avoid studying in front of the television and turn off the stereo. Many students insist they study better with background noise, and that may be true. Some students report good results with carefully selected and controlled music. The overwhelming majority of research indicates that silence is the best form of music for study.

At times noise may seem out of your control. A neighbor or roommate decides to find out how far he can turn up his stereo before the walls crumble. Meanwhile, your concentration on the principles of sociology goes down the tubes.

To get past this barrier, schedule study sessions for times when your living environment is usually quiet. If you live in a residence hall, ask if study rooms are available. Or go somewhere else, where it's quiet, such as the library. Some students have even found refuge in quiet restaurants, laundromats, and churches.

15. Notice how others misuse your time. Be aware of repeat offenders. Ask yourself if there are certain friends or relatives who consistently interrupt your study time. If avoiding the interrupter is impractical, send a clear message. Sometimes others don't realize they are breaking your concentration. You can give them a gentle yet firm reminder. If your message doesn't work, there are ways to make it more effective. For more ideas, see Chapter Ten.

Things You Can Ask Yourself When You Get Stuck

16. Ask: What is one task I can accomplish toward my goal? This is a useful technique to use on big, imposing jobs. Pick out one small accomplishment, preferably one you can complete in about five minutes; then do it. The satisfaction of getting one thing done often spurs you on to get one more thing done. Meanwhile, the job gets smaller.

17. Ask: Am I being too hard on myself? If you are feeling frustrated with a reading assignment, noticing that your attention wanders repeatedly, or falling behind on problems due for tomorrow, take a minute to listen to the messages you are giving yourself. Are you scolding yourself too harshly? Lighten up. Allow yourself to feel a little foolish and get on with it. Don't add to the problem by berating yourself.

Worrying about the future is another way people beat themselves up: "How will I ever get this all done?" "What if every paper I write turns out to be this hard?" "If I can't do the simple calculations now, how will I ever pass the final?" Instead of promoting learning, such questions fuel anxiety.

Labeling and generalizing weaknesses are other ways people are hard on themselves. Being objective and specific will eliminate this form of self-punishment and will likely generate new possibilities. An alternative to saying "I'm terrible in algebra" is to say "I don't understand factoring equations." This suggests a plan to improve.

18. Ask: Is this a piano? Carpenters who build rough frames for buildings have a saying they use when they bend a nail or hack a chunk out of a two-by-four: "Well, this ain't no piano." It means perfection is not necessary.

Ask yourself if what you are doing needs to be perfect. You don't have to apply the same standards of grammar to review notes that you apply to a term paper. If you can complete a job 95 percent perfectly in two hours, and 100 percent perfectly in four hours, ask yourself whether the additional 5 percent improvement is worth doubling the amount of time you spend.

Sometimes it *is* a piano. A tiny mistake can ruin an entire lab experiment. Computers are notorious for turning little errors into monsters. Accept lower standards only where they are appropriate.

A related suggestion is to weed out low-priority tasks. The to-do list for a large project can include dozens of items. Not all of them are equally important. Some can be done later on, and others could be skipped altogether if time is short.

Apply this idea when you study. In a long reading assignment, look for pages you can skim or skip. When it's appropriate, read chapter summaries or article abstracts. When reviewing your notes, look for material that may not be covered on a test and decide whether you want to study it.

19. Ask: Would I pay myself for what I'm doing right now? If you were employed as a student, would you be earning your wages? Ask yourself this question when you notice that you've taken your third popcorn break in 30 minutes. Most students are, in fact, employed as students. They are investing in their own productivity and paying a big price for the privilege of being a student. Sometimes they don't realize what a mediocre job may cost them.

20. Ask: Can I do just one more thing? Ask yourself this question at the end of a long day. Almost always you will have enough energy to do just one more short task. The overall increase in your productivity might surprise you.

21. Ask: Am I making time for things that are important but not urgent? If we spend most of our time putting out fires, we may feel drained and frustrated. According to Stephen R. Covey, this happens when we forget to take time for things that are fully important but not urgent. Examples are regular exercise, reading, prayer or meditation, quality time with friends and family, solitude, traveling, and cooking nutritious meals. Each of these can contribute directly to a long-term goal or life mission. Yet when schedules get tight, it's tempting to let these things go for that elusive day when we'll "finally have more time."

That won't come until we choose to make time for what's truly important. Knowing this, we can use some of the suggestions in this chapter to free up more time.

22. Ask: Can I delegate this? Instead of slogging through complicated tasks alone, you can draw on the talent and energy of other people. Busy executives know the value of delegating tasks to coworkers. Without delegation, many projects would flounder or die.

You can apply the same principle. Instead of doing all the housework or cooking by yourself, for example, assign some of the tasks to family members or roommates. Rather than making a trip to the library to look up a simple fact, call and ask a library assistant to do it. Instead of driving across town to deliver a package, hire a delivery service to do it. All these tactics can free up extra hours for studying.

It's not practical to delegate certain study tasks, such as writing term papers or completing reading assignments. However, you can still draw on the ideas of other people in completing such tasks. For instance, form a writing group to edit and critique papers, brainstorm topics or titles, and develop lists of sources.

If you're absent from a class, find a classmate to explain the lecture, discussion, and any assignments due. Presidents depend on briefings. You can use the technique too.

23. Ask: How did I just waste time? Notice when time passes and you haven't accomplished what you planned. Take a minute to review your actions and note the specific ways you wasted time. We operate by habit and tend to waste time in the same ways over and over again. When you are aware of things you do that kill your time, you are more likely to catch yourself in the act next time. Observing one small quirk may save you hours. One reminder: Noting how you waste time is not the same as feeling guilty about it. The point is not to blame yourself but to increase your skill. That means getting specific information about how you use time.

24. Ask: Could I find the time if I really wanted to? Often the way people speak rules out the option of finding more time. An alternative is to speak about time with more possibility.

The next time you're tempted to say, "I just don't have time," pause for a minute. Question the truth of this statement. Could you find four more hours this week for studying? Suppose that someone offered to pay you $10,000 to find those four hours. Suppose, too, that you will get paid only if you don't lose sleep, call in sick for work, or sacrifice anything important to you. Could you find the time if vast sums of money were involved?

Remember that when it comes to school, vast sums of money *are* involved.

25. Ask: Am I willing to promise it? This may be the most powerful time management idea of all. If you want to find time for a task, promise yourself—and others—that you'll get it done. To make this technique work, do more than say that you'll try or that you'll give it your best shot. Take an oath, as you would in court. Give your word.

One way to accomplish big things in life is to make big promises. There's little reward in promising what's safe or predictable. No athlete promises to place seventh in the Olympics. Chances are that if we're not making large promises, we're not stretching ourselves.

The point of making a promise is not to chain ourselves to rigid schedules or impossible expectations. We can also promise to reach goals without unbearable stress. We can keep schedules flexible and carry out our plans with ease and satisfaction.

At times we can go too far. Some promises are truly beyond us and we may break them. However, failing to keep a promise is just that—failing to keep a promise. A broken promise is not the worst thing in the world.

Promises can work magic. When our word is on the line, it's possible to discover reserves of time and energy we didn't know existed. Promises can push us to a breakthrough.

Keep going?

Some people keep going, even when they get stuck or fail again and again. To such people belongs the world. Consider the hapless politician who compiled this record: failed in business 1831; defeated for legislature 1832; second failure in business 1833; suffered nervous breakdown 1836, defeated for Speaker 1838; defeated for Elector 1840; defeated for Congress 1843; defeated for Senate 1844; defeated for Vice President 1856; defeated for Senate 1858; elected President 1860. Who was the fool who kept on going in spite of so many failures? Answer: *The fool was Abraham Lincoln.*

Student-Leader Relationships

Kristin Ganschow, workshop leader

The Problem:

A problem often faced by workshop leaders is maintaining a student-leader relationship that conveys friendship and encourages group interaction, while at the same time maintaining some sort of authority and control over the group. Often the students will look to the leader for all the answers, which can detract from group learning and interaction. The workshop leader must somehow step down from the teacher role, but still keep enough control to head the group in the right direction and enough authority to obtain and keep the students' trust.

Possible Causes:

A possible cause of this problem is the leaders' desire to teach the students everything they know about organic chemistry. It is often easy to step in too soon and show students the answer rather than letting the group work things out for themselves. Watching others make mistakes is difficult when the answer seems obvious. A cause for the opposite problem—a lack of authority—can be a lack of preparation. If the leader does not know what a problem is all about or makes too many mistakes, the students' trust could be lost.

Possible Impact:

If left unattended, an overly ambitious workshop leader could wind up handing out information in answer-key format rather than letting students discover and truly understand the concepts behind the problems. The workshops will turn into a large lecture format where the leader is spewing out information that the students will try to ingest. Conversely, if a leader has no authority, the students may not respond to the group leader's directions and end up in the middle of Kansas looking for the Sears Tower.

Suggestions:

Here are several things workshop leaders should keep in mind to achieve a good student-leader relationship:

- Be honest with students. Tell them you are not there to give them answers, but to facilitate group interaction.
- Direct questions elsewhere. When students look to you for help, try to get them to help each other. Do not step in until everyone seems clueless.
- Rely on confidence and knowledge to convey authority. Even if you infrequently answer questions and insist you don't have all the answers, a confidently answered question once and awhile will keep the students' trust.

Following these guidelines will not only help you to achieve a good balance in the leadership position, but will also foster good group interaction as well.

Dealing with a Question that Stumps the Leader: Finding the Answer

Laurie Kurtelawicz, workshop leader

One of the things both new and experienced workshop leaders fear is not knowing the answer to a problem or question. Most people will encounter this circumstance sooner or later, and contrary to popular belief, it probably won't cast doubt on your worthiness to be a workshop leader.

The most important thing to remember when you are in this situation is: Don't Fake It!! The group will know if you are making an answer up and that will cause them to lose confidence in you.

The next most important thing to remember is Don't Panic!! As a workshop leader, you have many resources available that can help you find the answer to a tough problem. One resource is the textbook. It can help you to understand the problem. After you work through the problem with your group, it often helps students if you tell them the area in the text where they can review the principles behind the problem. Other workshop leaders are also an excellent resource. Doing the problems ahead of time with other workshop leaders gives you a chance to pool information with others when dealing with difficult problems. It also offers you the opportunity to practice explaining the reasoning behind the problems. Another resource which is often overlooked by workshop leaders is the professor of the class. Don't be afraid to ask him or her questions if you are really stuck.

One of the most constructive things you can do in the workshop itself is to direct the question back at the group. Ask questions such as, "Does anybody have any ideas on this topic?" More often than not, together the group will be able to solve the problem using their collective knowledge. You can lessen the stress of this situation before it happens by reminding your students that, while you are qualified to help them, you aren't a professor and don't have all of the answers.

How Do I Get My Students to Work Together? Getting Cooperative Learning Started*

Marcy Hamby Towns
Department of Chemistry, Ball State University, Muncie, IN 47306-0445

When educating future scientists we often find ourselves lecturing to our students and encouraging them to perform tasks in a competitive individualistic environment. We ask students to work by themselves, set up reward structures so that cooperation among students is discouraged, emphasize technical competence to the exclusion of all else, and promote or pass students who produce little acceptable work. This does not represent the world in which they will be asked to work and interact.

The American Chemical Society Committee on Professional Training presented in their Spring 1996 newsletter the results of an industrial roundtable that was convened to address what industry looks for in new hires (1). Roundtable participants voiced broad agreement that in addition to technical skills, one of the key experiences industry seeks in new hires is team problem-solving. Chemists must be comfortable working with a diversity of people inside and outside their organization. Working on multidisciplinary teams, across departments, and between companies is becoming more prevalent. Modern science, in the academic setting, also increasingly requires teams of people working together to effectively solve problems. Emerging or growing fields such as biotechnology and materials science require teams of people with different areas of technical expertise.

Many undergraduate chemistry programs do not specifically incorporate techniques that promote students working in teams. Often chemistry students are expected to work individually, and working together is considered suspect (1,2). At the undergraduate level we must include experiences that train chemistry students how to work effectively with other students. The interpersonal skills developed while working in groups may be the set of skills most important to a chemist's employability, productivity, and career success (3,4).

In response to this problem, colleges have begun exploring ways to improve student learning and team problem-solving skills. Cooperative learning activities promote the development of interpersonal skills and communication skills through face-to-face interactions. During cooperative learning activities a group of students create an environment where they actively engage in the material by sharing insights and ideas, providing feedback, and teaching each other. Major reviews of cooperative learning research (5,6) have shown that cooperative learning leads to higher achievement, increased positive attitudes toward the subject area studied, higher self-esteem, greater acceptance of differences among peers, and enhanced conceptual development in a wide range of settings and across content areas.

Implementing Cooperative Learning

Some faculty discover cooperative learning through workshops or reading journals (7) and become excited about integrating a new technique into their classroom. They put their students into groups, turn them loose, and wait to see the aforementioned

* *J. Chem. Ed.* **1998**, *75*, 67-69

increase in achievement, increased positive attitudes, etc., and become disenchanted and disappointed when those outcomes do not occur.

Implementing cooperative learning requires preparation. Simply placing the students into groups and telling them to work together is not enough because it invokes two fallacies. One, that students actually know how to work together, and two, that students who do know how to work together will actually do so. So the question *before* you implement cooperative learning is, how do you get your students to work together?

Answers can be found in two texts that are helpful for teachers, especially chemistry teachers, who are trying to understand the principles and applications of cooperative learning. Johnson and Johnson have written dozens of texts and articles on cooperative learning; however, one text focuses exclusively on cooperative learning in the college classroom (8). This text contrasts the perspective of students as passive recipients of knowledge with the perspective of knowledge as constructed in the mind of the learner, reviews cooperative learning research, and outlines the essential elements of cooperative learning. It also provides many applications of different types of cooperative learning activities. A second valuable text, compiled and edited by Nurrenberg (9), is directed towards chemistry instructors who wish to incorporate cooperative learning activities into their classrooms. This text discusses what cooperative learning is and why it should be employed, the role of the teacher within a cooperative classroom, how to design tasks, and how to manage groups, and it addresses the common concerns many practitioners of cooperative learning face. Accompanying the text are field-tested examples of cooperative learning activities contributed by chemistry middle school, high school, and college instructors. These activities can be adapted to different classrooms and can serve as a catalyst for the development of new activities.

Reading these two texts helped me generate the materials and the approach I now use with my physical chemistry students. At this point, many articles have shaped the way I implement and use cooperative learning in my classroom. However, if I had to recommend a place to start reading about cooperative learning and thinking about how to get students to work together, I would strongly recommend these two texts.

How Do I Get My Students to Work Together?

The first day of class I talk to my students about cooperative learning and the activities they will be engaged in during the semester. I discuss with them why I think it is important to learn to work with other people, and I emphasize that they will learn more by explaining their understanding of concepts and problems to fellow students. From experience, I have found that it is important to do activities that help students in each group to get to know each other and help them craft the operating rules for their groups. The handouts and activities described in this article represent the present incarnation of my course.

Getting to Know You

The first day of class my students fill out a getting-to-know-you questionnaire, which identifies their intended career paths, previous cooperative learning experience, and attitudes towards group work. I collect the questionnaires and form the students into heterogeneous groups based on their responses, academic ability, and gender. It is very important that the instructor form the groups. If the students suggest choosing their own groups (any student who has had a negative experience with cooperative learning might), then take the opportunity to explain why you are forming the groups. Students need to recognize that in the workplace they will be expected to get along with many different types of people. The workplace of the 21st century will be composed of more women and minorities than ever before. In order to succeed in this type of workplace, students must learn to value and respect diversity—they need to know how to get along with different types of people. Cooperative learning activities can provide the first step in this direction.

Individual and Group Responsibilities Yield Group Covenants

In preparation for cooperative learning activities I want students to think about what they expect from their group and what they expect to contribute to their group. To achieve that end, the students are asked to respond to the following four questions before their first cooperative learning activity: (i) list your responsibilities to the group, (ii) list the responsibilities the group has to each member, (iii) describe the advantages of working in a group or as a team, and (iv) describe the disadvantages of working in a group or as a team (9). The students are asked to bring their written answers to the first cooperative learning activity held during the second day of classes.

During their first cooperative learning activity the students assemble in their assigned groups to compare and discuss their responses to each of the above questions. This enables the group to draft "Group Covenants," which delineate the individual group members' responsibilities to the group and the group's responsibilities to each member (9). In essence, this sets up the operating rules for each group. It helps to define acceptable behavior and how group members are to deal with each other in order to get work done. Each group member keeps a copy of the group covenants, and one copy signed by each group member is delivered to me.

Looks Like, Sounds Like

During the first cooperative learning activity, I direct a class discussion to compare what each group has written as individual responsibilities and group responsibilities in their group covenants. As each group states their group covenants, I record these responsibilities on the board. We find that most groups list similar responsibilities. They expect individuals to arrive prepared and on time, to contribute to the group, to listen to each other, to ask questions, and to be patient and respective of each other. They expect the group to work together to solve problems, to help each member understand how to solve problems, to encourage each other, to be open to learning from other members, to be respectful and patient with each other, and to make sure everyone understands.

After all the groups have chimed in, I ask the students to operationalize these behaviors by generating a response to the general question "What does each of these responsibilities look like and sound like?" (8). This is not a trivial activity. Students need to know that verbal and nonverbal behavior play a role in how people perceive them. For example, in my classroom groups often cite listening to other group members as a responsibility of each individual member to the group. But when asked what does listening "look like," the students appear to be rather perplexed. They ask what do I mean, "look like"? I explain that listening to a person involves actually looking at that person, making eye contact, and holding a facial expression that communicates that you are listening. Students need to realize that behaviors such as not looking at a person who is speaking, doodling, rolling their eyes, or sighing instantly telegraphs a message to the speaker. It declares that what the speaker is saying is not important. Asking students what their group covenants look like and sound like forces the students to think about how they are going to put their group covenants into practice.

Salaries and the Real World

As a final step on the first day of cooperative learning activities I talk about the starting salaries for B.S., M.S. and Ph.D. chemists as reported in the most recent yearly survey of chemists in *Chemical and Engineering News* (10). I emphasize that many of the most interesting and high-paying jobs in the workplace involve getting people to cooperate, leading others, and coping with complex power issues (3,8). Thus, the interpersonal and communication skills that the students develop during group work have a profound effect on their future success in the workplace.

How Are We Doing?

Groups need to regularly reflect on how well they are functioning. This feedback loop is known as group processing. It is critical to enhancing group performance and is analogous to evaluation and team-building efforts that occur in industrial settings (8,11). Processing can be facilitated by an evaluation form on which each team member evaluates others on the criteria outlined in their group covenants and completes the following statements.

1. To operate as an effective team we need to continue to do the following things:

2. To operate as a more effective team we need to start doing the following things:

3. To operate as a more effective team we need to stop doing the following things:

4. To carry out these actions here's what we are going to do (What's your strategy to address the concerns raised in the first three statements?):

The data generated from this activity can be used in two different fashions. First, faculty can use the individual evaluations as a check to make sure that students are functioning effectively in their group. If that is not the case, then action can be taken (easily done if the evaluation scores are factored into the student's grade). Second, by completing the four statements noted above each group can reflect on and discuss what things they do well and what they need to do better to function more effectively. This discourse reminds group members that building an effective group is not a static endeavor; it is a dynamic process, which requires vigilance and effort.

Building Classroom Community

Cooperative learning has many desirable outcomes (12-14). It helps students build a feeling of community in the classroom and fosters a warmer classroom climate, which promotes learning and achievement. This warmer climate expresses itself in the students forming friendships and challenging and encouraging each other to truly understand the material. Cooperative learning activities encourage students to engage in the type of discourse about concepts and problem solving that moves them away from rote learning strategies and toward more meaningful learning strategies. Students strive to understand different ways of explaining concepts and different perspectives on solving problems. When students experience a feeling of community they become more willing to take on tough tasks because they expect to succeed; absenteeism drops; and their attitude toward the course becomes more positive. Their potential for achievement becomes enormous.

Building this community and realizing this potential require the students know and trust each other. From the faculty's point of view, this implies that facilitating student learning during cooperative learning activities begins by helping students learn how to work with each other. Performing the activities described in this article will help the students get to know each other, create operating rules for their groups, define and describe these rules, and evaluate and improve how their group functions. Groups that function effectively and thus build a sense of community unlock their potential to achieve and perform at a high level.

Literature Cited

1. The American Chemical Society Committee on Professional Training, *CPT Newsletter* **1996**, *2*, 5.
2. Rayman, P.; Brett, B. *Pathways for Women in the Sciences*; Pathways Project, Wellesley College Center for Research on Women: Wellesley, MA, **1993**.
3. Johnson, D. W.; Johnson, R. T. *Ed. Leadership* **1989**, *47*, 29-33.
4. Tobias, S.; Chubin, D. E.; Aylesworth, K. *Rethinking Science as a Career: Perceptions and Realities in the Physical Sciences*; Research Corporation: Tucson, AZ, **1995**.
5. Cohen, E. G. *Rev. Ed. Res.* **1994**, *64*, 1-35.
6. Qin, A.; Johnson, D. W.; Johnson, R. T. *Rev. Ed. Res.* **1995**, *65*, 129-143.
7. Nurrenbern, S. C.; Robinson, W. R. *J. Chem. Educ.* **1997**, *74 , 623-624.*
8. Johnson, D. W.; Johnson, R. T.; Smith, K. A. *Active Learning: Cooperation in the College Classroom*; Interaction Book Co.: Edina, MN, **1991**.
9. *Experiences in Cooperative Learning: A Collection for Chemistry Teachers*; Nurrenbern, S.C., Compiler, Ed.; University of Wisconsin-Madison: Madison, WI, *1995*, ICE Pub. 95-001.
10. Heylen, M. *Chem. Eng. News* **1997**, *73(32)*, 12-15.
11. Manning, G.; Curtis, K.; McMillan, S. *Building Community: The Human Side of Work*; Thomson Executive Press: Cincinnati, OH, **1996**.
12. Kreke, K.; Towns, M. H. Student Perspectives of Cooperative Learning Activities; submitted to *J. Res. Sci. Teach.*
13. Towns, M. H.; Grant, E. R.; I Believe I Will Go Out of This Class Actually Knowing Something: Cooperative Learning Activities in Physical Chemistry; *J. Res. Sci. Teach.* **1997**, *34*, 819-835.
14. Kreke, K.; Towns, M. H. Cooperative Learning Activities in Physical Chemistry: It's Nice to Finally Realize that the World Needs People with Social Skills as well as Knowledge. Presented at 14th Biennial Conference on Chemical Education, August **1996**, Clemson, SC.

The First Day

Jeremy Kukafka, workshop leader

The first workshop session is the most important workshop of the semester. It is where the students are first introduced to the "workshop" format, the workshop leader, each other, and to organic chemistry problem solving. The recipe for a successful workshop requires a successful first day. All too often, critical issues are not stressed, and the group is sent down the path of the all-too-boring silent, ineffective workshop.

When students enter the workshop environment for the first time, they do not know what to expect. They have already been through traditional discussion sections and recitations, and most of them will initially feel more comfortable in that setting than in the workshop environment. As a result, it would not be unusual for them to use the recitation style as a safety blanket when first entering the workshop program. This is what we do NOT want to happen. A successful first day, in order to bring students out of the realm of the recitation and into the realm of the workshop, must do the following:

A. Give students a general understanding of the workshop program. They should understand the goals of the program, and what it requires of them as students.

B. Show the students what workshop leaders are. They're not scary. They are just normal students who want to help everyone in their workshop understand the material in the most enjoyable and effective way possible. Ideally, this would lead everyone to an "A" in the course. Let it be known the leader is there to help.

C. Create a pleasant atmosphere where learning is both effective and fun. Give students some time to get to know one another. Avoid the following: "My name is __________, and I am a ____________ major." It produces the aura of a traditional classroom.

   BIG HELP (from experience): If the leader appears goofy and imperfect, then the students will be more comfortable being themselves.

D. Give the students a SEMI-format for solving workshop problems. No rigid structure should be given. What should be stressed is that the students should be working together actively. The more chem-, bio-, physics-, or math-noise, the better.

   If these suggestions are followed, and the leader adds a little bit of his/her "special spice," then there's no doubt the workshop will be off to a successful start.

Helping a Group That Won't Talk Much

Ryan Rekuski, workshop leader

Silence. At times, nothing can be sweeter. So quiet you can hear your own heart beat. However, your workshop is neither the time nor the place. After all, it's not a test. It's an adventure in group learning, a collaborative effort to help each other learn and understand the subject. What you want is lively discussion and interaction.

What if this does not happen? What if the group sits quietly and stares at you, waiting for you to spit out answers like the average T.A.? What if they work, but by themselves, like it really were an exam? What if they only talk to the one other person in the workshop they know? What if there is dead silence? What is a workshop leader to do? How does one facilitate interaction and discussion?

Well, let's look at some possible causes, the effect they can have on the workshop, and finally some viable solutions.

1. *It is early in the semester and people are still apprehensive about the workshop and don't feel comfortable with the other members in the group and the leader.*

If a level of comfort in which people feel they can freely ask or answer questions is never established, people will never open up. They will sit quietly and wait for answers to be given by someone else. Group members hope that person will be you. There will be neither interaction nor discussion. The group will get accustomed to the format of this "faux workshop," which will make it hard to develop a good group dynamic as the semester progresses.

This problem can be alleviated by establishing a high level of comfort early. Get students talking to each other right away, even if it is not always about the course content. Give them time in the beginning to talk and set up relationships. It will help them become more comfortable, which will lead to better group dynamics, better discussion and a better overall experience for everyone.

2. *The group is just quiet and reserved, introverted.*

This is not really a "problem." It can lead to some of the same problems as in (1), in that there could be a lack of interaction and discussion. However, this may not be indicative of a lack of learning and benefit. People may benefit from being able to talk to others, when necessary, though it may not be often.

There is not much to "solve" here. Try to get students to interact, but allow them to have their space. Let them work in a way that is best for the group. If it's quiet, that's okay, as long as the workshop is productive.

3. *Students come unprepared.*

This is a huge problem. If the cause of the silence is unprepared students, a workshop becomes an exercise in futility. Unprepared students slow the group down, and want you to teach them the material. Those students who do come prepared begin to resent those who do not. People feel like the workshop is not helpful. It is frustrating for the leader as well, because this can cause students' expectations of the leader to rise. They expect you to have answers when they should be working together to find them. You become the answer key, not a resource.

To resolve this, try to find out why they are unprepared. Then, let students know early that the material builds on itself. Understanding early concepts is vital to comprehending later material. Tell them about personal experiences of how being prepared/unprepared helped/hurt your performance. Let them know the implications of their actions, how their being unprepared hurts the group.

If the cause of poor preparation is that students feel lost and confused, point them in the right direction to get help. Remind them that the professor has office hours for exactly those reasons. Tell them about the learning center, study groups, and tutoring. Be the resource of information that you are.

How Do I Handle the Student Who Gets Everything Right? Dealing with Uneven Abilities in the Workshop

Andrew Epstein, workshop leader

A potential problem in the dynamics of groups is an uneven distribution in the intrinsic abilities of the members. While intelligence/academic performance is by no means an "all-or-none" phenomenon, and certainly relies heavily on steady work habits, interest in learning, and more personal motivations (vocational, etc.), some individuals may be "natural" or more comfortable with certain material. This heterogeneity is particularly believed to be present in scientific education, in terms of the stereotypical "math and science" person—these people just seem to "get it." Let's say we have such an individual in our workshop—what might this mean for the rest of the group? A workshop leader doesn't want to restrain the maximal advancement of his/her better students, nor does he/she want to ignore the needs of other students who must have their skills enhanced at a different pace!

This set of circumstances can worsen very quickly if left unattended. Students who excel may not be immediately interested in sharing their talent, may not have the confidence to do so, or may not know how. Other less advanced students may become frustrated with their own efforts, and resent the "ease" with which fellow group members reach an understanding of concepts. These people may lose confidence in themselves, withdraw from the group in terms of participation and contribution, and eventually not even care about the subject in any way—all problems that the cooperative learning method in workshop attempts to extinguish. In light of these thoughts, the following are some ideas that may help your group dynamics:

- Pair strong students with weaker students in group activities. This will encourage strong students to share their thought processes in a closer setting. Weaker students may be less inclined towards embarrassment if fewer people are watching. They may also develop a friendly working relationship with the stronger student that will enhance confidence, interest, and motivation. Presentations to the group are made by the weaker student with the assurance of the stronger student's input—a potentially huge confidence builder that improves participation.

- After seeing this happen, try rotating your pairs. This gradually challenges your weaker students to take some confidence and individual thinking skills with them to a new mini-group. Try to maintain the stronger-weaker ratio as well as you can.

- Chances are your stronger students are aware of their abilities relative to the group. Take them aside and praise them on their talent, and challenge them to integrate their abilities. Make mention of how they are a valuable resource—remind them this is THEIR group. Everyone wants to feel important!

- Remember to let students work individually on occasion as well. This is a great opportunity to challenge the intellects of your future leaders with more advanced concepts. Push them! I always enjoyed figuring out what stumped me. This individual work also provides weaker students with a chance to actualize their improvement. Tailor the problems you give them to their current abilities, but don't necessarily baby them. When they get something right on their own, it is invaluable for growth.

As always, a successful workshop is a unique combination of these ideas. Good luck, leaders!

Reaching the Second Tier—Learning and Teaching Styles in College Science Education*

Richard M. Felder†

In her recent study of college science instruction, Sheila Tobias (1990) defines two tiers of entering college students: the first consisting of those who go on to earn science degrees; and the second, those who have the initial intention and the ability to do so but instead switch to nonscientific fields. The number of students in the second category might in fact be enough to prevent the shortfall of American scientists and engineers that has been widely forecast for the coming decade.

The thrust of Tobias's study is that introductory science courses are responsible for driving off many students in the second tier. The negative features of the courses she cites include the following:

- Failure to motivate interest in science by establishing its relevance to the students' lives and personal interests

- Relegation of students to almost complete passivity in the classroom

- Emphasis on competition for grades rather than cooperative learning

- Focus on algorithmic problem solving as opposed to conceptual understanding

Recent educational research provides theoretical support for Tobias's assertions, which are based largely on anecdotal accounts. The research shows that students are characterized by significantly different *learning styles*: they preferentially focus on different types of information, tend to operate on perceived information in different ways, and achieve understanding at different rates (Barbe and Milone 1991; Claxton and Murrell 1987; Corno and Snow 1986; Felder 1988, 1989, 1990; Felder and Silverman 1988; Godleski 1984; Kolb 1984; Lawrence 1982; Pask 1988; Schmeck 1988). Students whose learning styles are compatible with the teaching style of a course instructor tend to retain information longer, apply it more effectively, and have more positive post-course attitudes toward the subject than do their counterparts who experience learning/teaching style mismatches. Most of the points raised by Tobias about the poor quality of introductory college science instruction can be expressed directly as failures to address certain common learning styles.

Felder and Silverman (1988) have synthesized findings from a number of studies to formulate a learning-style model with dimensions that should be particularly relevant to science education. In the sections that follow, the model dimensions are briefly summarized and instructional methods are then proposed that should reach students who span the spectrum of learning styles, including students in Tobias's second tier.

---

* J. College Sci. Teaching **1993**, 286-290.

† Richard M. Felder is Hoechst Celanese Professor of Chemical Engineering at North Carolina State University, Raleigh, NC 27695-7905.

Dimensions of Learning Style

A student's learning style may be defined in part by the answers to five questions:

1. What type of information does the student preferentially perceive: *sensory* –sights, sounds, physical sensations, or *intuitive*—memories, ideas, insights?

2. Through which modality is sensory information most effectively perceived: *visual*—pictures, diagrams, graphs, demonstrations, or *verbal*—sounds , written and spoken words and formulas?

3. With which organization of information is the student most comfortable: *inductive*—facts and observations are given, underlying principles are inferred, or *deductive*—principles are given, consequences and applications are deduced?

4. How does the student prefer to process information: *actively*—through engagement in physical activity or discussion, or *reflectively*—through introspection?

5. How does the student progress understanding: *sequentially*—in a logical progression of small incremental steps, or *globally* - in large jumps, holistically?

The dichotomous learning-style dimensions of this model (sensing/intuitive, visual/verbal, inductive/deductive, active/reflective, and sequential/global) are continuous and not either/or categories. A student's preference on a given scale (for example, for inductive or deductive presentation) may be strong, moderate, or almost nonexistent, may change with time, and may vary from one subject or learning environment to another.

Sensing and Intuitive Perception

People are constantly being bombarded with information, both through their senses and from their subconscious minds. The volume of this information is much greater than they can consciously attend to: they therefore select a minute fraction of it to admit to their "working memory" and the rest of it is effectively lost. In making this selection, *sensing learners* (sensors) favor information that comes in through their senses and *intuitive learners* (intuitors) favor information that arises internally through memory, reflection, and imagination. These categories derive from Carl Jung's theory of psychological types. The strength of an individual's preference for sensation or intuition can be assessed with the *Myers-Briggs Type Indicator* (Lawrence 1982; Myers and McCaulley 1986). [For a further discussion of the MBTI see the "Research & Teaching" column, starting on p. 276 of this *JCST* issue.]

Sensors tend to be practical; intuitors tend to be imaginative. Sensors like facts and observations; intuitors prefer concepts and interpretations. A student who complains about courses having nothing to do with the real world is almost certainly a sensor. Sensors like to solve problems using well-established procedures, don't mind detail work, and don't like unexpected twists or complications; intuitors like variety in their work, don't mind complexity, and get bored with too much detail and repetition. Sensors tend to be careful but may be slow; intuitors tend to be quick but may be careless (Felder 1989).

Sensing learners learn best when given facts and procedures, but most science courses (particularly physics and chemistry) focus on abstract concepts, theories, and formulas, putting sensors at a distinct disadvantage. Moreover, sensors are less comfortable than intuitors with symbols; since words and algebraic variables - the stuff of examinations—are symbolic, sensors must translate them into concrete mental images in order to understand them. This process can be a lengthy one, and many sensors who know the material often run out of time on tests. The net result is that sensors tend to get lower grades than intuitors in lecture courses (Godleski 1984); in effect, they are selectively weeded out, even though they are as likely as intuitors to succeed in science and engineering careers (Felder 1989).

Visual and Verbal Input

Visual learners get more information from visual images (pictures, diagrams, graphs, schematics, demonstrations) than from verbal material (written and spoken words and mathematical formulas), and vice versa for *verbal learners* (Bandler and Rindler 1979; Barbe and Milone 1982). If something is simply said and not shown to visual learners (for example, in a lecture), there is a good chance they will not retain it. Most people (at least in western cultures) and presumably most students in science classes are visual learners (Barbe and Milone 1981) while the information presented in almost every lecture course is overwhelmingly verbal—written words and formulas in texts and on the chalkboard, spoken words in lectures, with only an occasional diagram, chart, or demonstration breaking the pattern. Professors should not be surprised when many of their students cannot reproduce information that was presented to them not long before; it may have been expressed but it was never heard.

Inductive and Deductive Organization

Inductive learners prefer to learn a body of material by seeing specific cases first (observations, experimental results, numerical examples) and working up to governing principles and theories by inference; *deductive learners* prefer to begin with general principles and to deduce consequences and applications. Since deduction tends to be more concise and orderly than induction, students who prefer a highly structured presentation are likely to prefer a deductive approach while those who prefer less structure are more likely to favor induction.

Research shows that of these two approaches to education, induction promotes deeper learning and longer retention of information and gives students greater confidence in their problem-solving abilities (Felder and Silverman 1988; McKeachie 1980). The research notwithstanding, most college science instruction is exclusively deductive—probably because deductive presentations are easier to prepare and control and allow more rapid coverage of material. In the words of a student evaluating his introductory physics course, "The students are given premasticated information simply to mimic and apply to problems. Let them, rather, be exposed to conceptual problems, try to find solutions to them on their own, and then help them to understand the mistakes they make along the way" (Tobias 1990, p. 25). The approach suggested by this student is inductive learning.

Active and Reflective Processing

Active learners tend to learn while doing something active—trying things out, bouncing ideas off others; *reflective learners* do much more of their processing introspectively, thinking things through before trying them out (Kolb 1984). Active learners work well in groups; reflective learners prefer to work alone or in pairs. Unfortunately, most lecture classes do very little for either group: the active learners never get to do anything and the reflective learners never have time to reflect. Instead, both groups are kept busy trying to keep up with a constant barrage of verbiage, or else they are lulled into inattention by enforced passivity.

The research is quite clear on the question of active and reflective versus passive learning. In a number of studies comparing instructor-centered classes (lecture/demonstration) with student-centered classes (problem-solving/discussion), lectures were found to be marginally more effective when students were tested on short-term recall of facts but active classroom environments were superior when the criteria involved comprehension, long-term recall, general problem-solving ability, scientific attitude, and subsequent interest in the subject (McKeachie 1986). Substantial benefits are also cited for teaching methods that provide opportunities for reflection, such as giving students time in class to write brief summaries and formulate written questions about the material just covered (McKeachie 1986; Wilson 1986).

Sequential and Global Understanding

Sequential learners absorb information and acquire understanding of material in small connected chunks; *global learners* take in information in seemingly unconnected fragments and achieve understanding in large holistic leaps. Sequential learners can solve problems with incomplete understanding of the material and their solutions are generally orderly and easy to follow, but they may lack a grasp of the big picture - the broad context of a body of knowledge and its interrelationships with other subjects and disciplines. Global learners work in a more all-or-nothing fashion and may appear slow and do poorly on homework and tests until they grasp the total picture, but once they have it they can often see connections to other subjects that escape sequential learners (Pask 1988).

Before global learners can master the details of a subject they need to understand how the material being presented relates to their prior knowledge and experience, but only exceptional teachers routinely provide such broad perspectives on their subjects. In consequence, many global learners who have the potential to become outstanding creative researchers fall by the wayside because their mental processes do not allow them to keep up with the sequential pace of their science courses (Felder 1990).

Toward a Multistyle Approach to Science Education

Students whose learning styles fall in any of the given categories have the potential to be excellent scientists. The observant and methodical sensors, for example, make good experimentalists, and the insightful and imaginative intuitors make good theoreticians. Active learners are adept at administration and team-oriented project work; reflective learners do well at individual research and design. Sequential learners are often good analysts, skilled at solving convergent (single-answer) problems; global learners are often good synthesizers, able to apply material from several disciplines to solve problems that could not have been solved with conventional single-discipline approaches.

Unfortunately—in part because teachers tend to favor their own learning styles, in part because they instinctively teach the way they were taught in most college classes—the teaching style in most lecture courses tilts heavily toward the small percentage of college students who are intuitive, verbal, deductive, reflective, and sequential. This imbalance puts a sizeable fraction of the student population at a disadvantage. Laboratory courses, being inherently sensory, visual, and active, could in principle compensate for a portion of the imbalance; however, most labs involve primarily mechanical exercises that illustrate only a minor subset of the concepts presented in lecture and seldom provide significant insights or skill development. Sensing, visual, inductive, active, and global learners thus rarely get their educational needs met in science courses.

The mismatches between the prevailing teaching style in most science courses and the learning styles of most of the students have several serious consequences (Felder and Silverman 1988). Students who experience them feel as if they are being addressed in an unfamiliar foreign language; they tend to get lower grades than students whose learning styles are better matched to the instructor's teaching style (Godleski 1984) and are less likely to develop an interest in the course material (Felder 1988). If the mismatches are extreme, the students are apt to lose interest in science altogether and be among the more than 200,000 who switch to other fields each year after their first college science courses (Tobias 1990). Professors confronted by inattentive classes and poor student performance may become hostile toward the students (which aggravates the situation) or discouraged about their professional competence. Most seriously, society loses potentially excellent scientists.

These problems could be minimized and the quality of science education significantly enhanced if instructors modified their teaching styles to accommodate the learning styles of all the students in their classes. Granted, the prospect of trying to address 32^5 different learning styles simultaneously in a single class might seem forbidding to most instructors. The point, however, is not to determine each student's learning style and then teach to it exclusively but simply to address each side of each learning style dimension at least some of the time. If this balance could be achieved in science courses, the students would all be taught in a manner than sometimes matches their learning styles, thereby promoting effective learning and positive attitudes toward science, and sometimes compels them to exercise and hence strengthen their less-developed abilities, ultimately making them better scholars and scientists.

Major transformations in teaching style are not necessary to achieve the desired balance. Of the 10 defined learning style categories, five (intuitive, verbal, deductive, reflective, and sequential) are adequately covered by the traditional lecture-based teaching approach. Moreover, there is considerable overlap in teaching methods that address the style dimensions shortchanged by the traditional method (sensing, visual, inductive, active, and global). The systematic use of a small number of additional teaching methods in a class may therefore be sufficient to meet the needs of all of the students. Some of these methods are as follows:

- ***Motivate presentation of theoretical material with prior presentation of phenomena that the theory will help explain and problems that the theory will be used to solve*** *(sensing, inductive, global).* Don't jump directly into free body diagrams and force balances on the first day of the statics course. First, describe problems associated with the design of buildings and bridges and artificial limbs, and perhaps give the students some of those problems and see how far they can go with them before they get all the tools for solving them.

- ***Balance concrete information***—descriptions of physical phenomena, results from real and simulated experiments, demonstrations, and problem-solving algorithms (*sensing*)—***with conceptual information*** - theories, mathematical models, and material that emphasize fundamental understanding (*intuitive*)—***in all courses***. When covering concepts of vapor-liquid equilibria, go through Raoult's and Henry's law calculations and nonideal solution behavior . . . but also discuss the meaning of weather reports (the temperature is 27 °C, barometric pressure is 29.95 inches, and the relative humidity is 68%), the manufacture of carbonated beverages, and what you would observe if you poured 50 mL liquid benzene and 50 mL liquid toluene into an open flask, heated the flask, and monitored the liquid volume, temperature, and composition. Give the relations between torque, moments, and angular motion—but first get students to exert pressure on a door at different perpendicular distances from the hinges and then have them try to interpret the results.

- ***Make extensive use of sketches, plots, schematics, vector diagrams, computer graphics, and physical demonstrations*** (*visual*) ***in addition to oral and written explanations and derivations*** (*verbal*) ***in lectures and readings.*** Show flow charts of the reaction and transport processes that occur in particle accelerators, test tubes, and biological cells before presenting the relevant theories, and sketch or demonstrate the experiments used to validate the theories. "Look at this micrograph of a mammalian cell. Now here's a schematic showing the structures of the different organelles and their interrelations. OK, now let's consider individual organelle functions and how compartmentalization makes cell regulation and specialization possible."

- ***To illustrate abstract concepts or problem-solving algorithms, use at least some numerical examples*** (*sensing*) ***to supplement the usual algebraic examples*** (*intuitive*).

- ***Use physical analogies and demonstrations to illustrate the magnitudes of calculated quantities*** (*sensing, global*). "100 microns—that"s about the thickness of a sheet of paper." "Think of a mole as a very large dozen molecules." "Pick up this 100 mL bottle of water. Now pick up this 100 mL bottle of mercury. Now let's talk about density."

- ***Give some experimental observations before presenting the general principles and have the students (preferably working in groups) see how far they can get toward inferring the latter*** (*inductive*). Rather than giving the students Ohm's or Kirchoff's law up front and asking them to solve it for one unknown or another, give them experimental voltage/current/resistance data for several circuits and let them try to figure out the laws for themselves. Describe a situation in which a teakettle is placed on a stove burner and have the students estimate heat inputs and times required to boil and then completely vaporize the kettle contents, and them give them the necessary thermodynamic and mathematical tools and let them carry out the analysis rigorously (Felder 1991).

- ***Provide time in class for students to think about the material being presented*** (*reflective*) ***and for active student participation*** (*active*). Occasionally pause during a lecture to allow time for thinking and formulating questions. Assign "one-minute papers" close to the end of a lecture period, having students write on index cards the most important point made in the lecture and the single most pressing unanswered question (Wilson 1986). Assign brief group problem-solving exercises in class in which the students working in groups of three or four at their seats spend one or several minutes tackling any of a wide variety of questions and problems. ("Begin the solution to this problem." "Take the next step in the solution." "What's wrong with what I just wrote on the board?" "What assumptions are implicit in this result?" "Suppose you go into the laboratory, take measurements, and find that the formula we have just derived gives incorrect results: how many possible explanations can you come up with?")

- ***Encourage or mandate cooperation on homework*** *(active).* Students who participate in cooperative (team-based) learning experiences—both in and out of class—are reported to earn better grades, display more enthusiasm for their chosen field, and improve their chances for graduation in that field relative to their counterparts in more traditional competitive class settings (Cooper, Prescott, Cook, Smith, Mueck, and Cuseo 1990).

- ***Demonstrate the logical flow of individual course topics*** *(sequential),* ***but also point out connections between the current material and other relevant material in the same course, in other courses in the same discipline, in other disciplines, and in everyday experience*** *(global).* Before discussing cell metabolism chemistry in detail, describe energy release by glucose oxidation . . . and relate it to energy release by nuclear fission, electron orbit decay, waterfalls, and combustion in fireplaces, power plant boilers, and automobiles. Discuss where the energy comes from and where it goes in each of these processes and how cell metabolism differs from the other examples . . . and then consider the photosynthetic origins of the energy stored in the C-H bonds and the conditions under which the earth's supply of usable energy might eventually run out.

How can an instructor do all that and still get through the syllabus? One way is to put most of the material usually written on the board in handouts, go through the handouts quickly in class, and use the considerable class time saved for activities like those just suggested. The consequent gain in quantity and quality of the resulting learning will more than compensate for the photocopying costs.

A final suggestion is to talk to students about their learning styles, either in class or in advising. Many of them have been coping with mismatches between their learning style and their instructors' teaching styles since high school or earlier, attributing their difficulties to their own inadequacies. Telling struggling sensors or active or global learners in Sheila Tobias's second tier about their learning strengths, weaknesses, and educational needs may be the best way to get them to see for themselves that (in Tobias's phrase) "They're not dumb, they're different," and so to move some of them into the first tier, where they belong.

References

Bandler, R. and J. Grinder. 1979. *Frogs into Princes.* Moab, UT: Real People Press.

Barbe, W. D. and M. N. Milone. 1981. What we know about modality strengths. *Educational Leadership*, February: 378-380.

Claxton, C. S. and P. H. Murrell. 1987. *Learning Styles: Implications for Improving Educational Practice.* ASHE-ERIC Higher Education Report No. 4. ASHE, College Station.

Corno, L. and R. E. Snow. 1986. Adapting teaching to individual differences among learners. In M. Wittrock, Ed. *Handbook of Research on Teaching.* New York: Macmillan.

Cooper, J., S. Prescott, L. Cook, L. Smith, R. Mueck and J. Cuseo. 1990. *Cooperative Learning and College Instruction*. Long Beach, CA: California State University Foundation.

Felder, Richard. 1988. How students learn: Adapting teaching styles to learning styles. Proceedings, Frontiers in Education Conference. ASEE/IEEE, Santa Barbara, CA, p. 489.

_____. 1989. Meet your students: 1. Stan and Nathan. *Chemical Engineering Education*, Spring: 68.

_____. 1990. Meet your students: 2. Susan and Glenda. *Chemical Engineering Education*, Winter: 7.

_____. 1991. It goes without saying. *Chemical Engineering Education, Summer: 132.*

_____ and Linda Silverman. 1988. Learning and teaching styles in engineering education. *Engineering Education* 78(7): April: 674-681.

Godleski, Edward. 1984. Learning style compatibility of engineering students and faculty. Proceedings, Annual Frontiers in Education Conference. ASEE/IEEE, Philadelphia, p. 362.

Kolb, David. 1984. *Experiential learning: Experience as the Source of Learning and Development*. Englewood Cliffs, NJ: Prentice-Hall.

Lawrence, Gordon. 1982. *People Types and Tiger Stripes: A Practical Guide to Learning Styles*. 2nd Edition. Gainesville, FL: Center for Applications of Psychological Type.

McKeachie, Wilbert. 1980. Improving lectures by understanding students' information processing. In McKeachie, W. J., Ed., *Learning, Cognition, and College Teaching*. New Directions for Teaching and Learning, No. 2. San Francisco: Jossey-Bass, p. 32.

_____. 1986. *Teaching Tips: A Guidebook for the Beginning College Teacher*, 8th Edition. Lexington, MA: D. C. Heath & Co.

Myers, I. B. and M. H. McCaulley. 1986. *Manual: A Guide to the Development and Use of the Myers-Briggs Type Indicator*, 2nd Edition. Palo Alto: Consulting Psychologists Press.

Pask, G. 1988. Learning strategies, teaching strategies, and conceptual or learning style. In Schmeck, R., Ed., *Learning Strategies and Learning Styles*. New York: Plenum Press, Ch. 4.

Schmeck, Ronald, Ed. 1988. *Learning Strategies and Learning Styles*. New York: Plenum Press.

Tobias, S. 1986. Improving faculty teaching: Effective use of student evaluations and consultants. *Journal of Higher Education 57*: 196-211.

Are Our Students Conceptual Thinkers or Algorithmic Problem Solvers?

Identifying Conceptual Students in General Chemistry

Mary B. Nakhleh
Purdue University, West Lafayette, IN 47907-1393

Bright students who perceive the world somewhat differently from the traditional math-oriented science/engineering students sit in every large general chemistry lecture. These students desire to explore the why of chemistry more than the how of chemistry. That is, they are more interested in the concepts than in algorithmic problem solving.

Second-Tier Students in General Chemistry

Tobias (*1*) has studied these students and their reactions to general chemistry courses. She calls these students "second-tier students" and urges that some effort be made to recruit them into the study of chemistry.

I became interested in the problem of identifying these students who have the ability to study chemistry yet are not attracted to the discipline. This student population could be viewed as a potentially rich source of recruits into the scientific disciplines, which have experienced steady declines in majors for several years (*2*).

Testing for Conceptual Thinking and Problem-Solving Skills

However, no way has been devised of identifying, and thus recruiting, these students in the typically large lecture sections of general chemistry courses. This study created a short, simple test that might help identify these students in general chemistry by investigating differential performance on conceptual and problem-solving questions.

The test consists of five matched pairs of questions. Each pair deals with a specific area of chemistry. One question of each pair is phrased as an algorithmic problem-solving question, while the other question is phrased as a conceptual question whose solution requires understanding of the principles of the topic rather than an algorithm.

Constructing Pairs of Questions

I identified five areas of chemistry that were taught in each of the four general chemistry courses offered in the fall semester at a large midwestern university. I then constructed five pairs of questions so that each pair dealt with one specific area of general chemistry (see Figures 1–5).

pair 1: gas laws
pair 2: equations
pair 3: limiting reagents
pair 4: empirical formulas
pair 5: density

Within each pair, one question required the student to manipulate a formula or work through an algorithm to find a numerical solution to a problem. The second question of the pair required students to use their conceptual knowledge of the topic to select an answer. In pairs 1, 3, and 5 the conceptual questions required the students to interpret drawings. Conceptual questions in pairs 2 and 4 required students to interpret text only.

The test items were randomly incorporated into the final exams of the four first-semester freshman chemistry courses: remedial, science/engineering major, chemistry major, and honors. The bulk of the students were in the course for science and engineering majors. Approximately 1,000 students were involved in the study.

Hypotheses

There were three initial hypotheses.

> That the remedial course would contain the highest concentration of conceptual thinkers, rather than algorithmic problem solvers.
>
> That the honors students would demonstrate both conceptual thinking and algorithmic problem solving.
>
> That the science/engineering majors course would contain some conceptual thinkers but mostly algorithmic problem solvers.

1. 0.100 mole of hydrogen gas occupies 600 mL at 25 °C and 4.08 atm. If the volume is held constant, what will be the pressure of the sample of gas at –5 °C?

A. 4.54 atm B.* 3.67 atm C. 6.00 atm
D. 2.98 atm E. 4.08 atm

2. The following diagram represents a cross-sectional area of a rigid sealed steel tank filled with hydrogen gas at 20 °C and 3 atm pressure. The dots represent the distribution of all the hydrogen molecules in the tank.

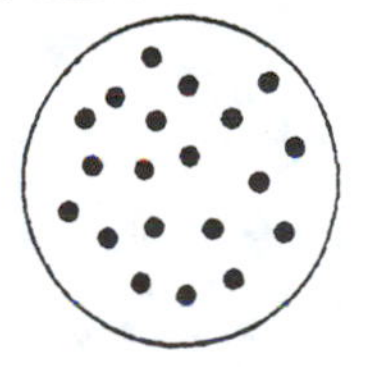

Which of the following diagrams illustrate one probable distribution of molecules of hydrogen gas in the sealed steel tank if the temperature is lowered to –5 °C? The boiling point of hydrogen is –252.8 °C.

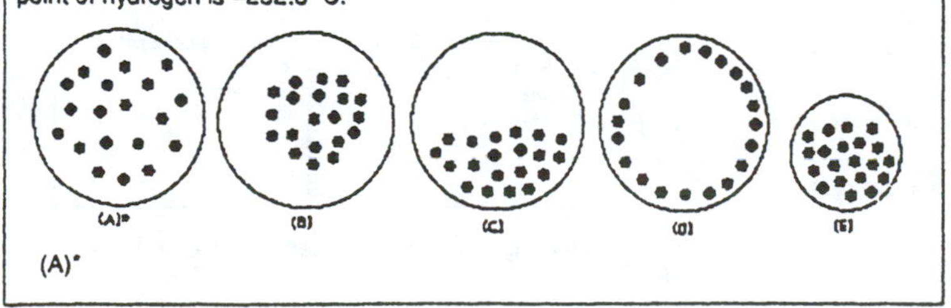

(A)*

Figure 1. Question pair for gas laws.

1. Calculate the maximum weight of NH_3 that could be produced from 1.9 mol of hydrogen and excess nitrogen according to the following reaction.

$$N_2 + 3H_2 \rightarrow 2NH_3$$

A. 15 g B. 28 g C.* 22 g D. 30 g E. 17 g

2. Any quantity of Cu in excess of one mole will always react with two moles of $AgNO_3$ to produce one mole of $Cu(NO_3)_2$ and two moles of Ag. Therefore we know that 1.5 moles of Cu will react with two moles of $AgNO_3$ to produce 215.74 grams of Ag. Which of the following concepts is the only concept NOT associated with these statements ?

A. Chemical reactions involve the rearrangement of atoms about one another.
B. In an ordinary chemical reaction mass is not created or destroyed.
C. Identical compounds are always composed of the same elements in the same proportion by mass.
D.* NO_3^- is easily reduced.
E. The number of moles of products formed in this case are determined by the number of grams of $AgNO_3$ available.

Figure 2. Question pair for equations.

1. Which is the limiting reagent when 2.0 mol of CO_2 reacts with 2.0 mole of S_2 to form COS and O_2?

A.* CO_2 B. S_2 C. COS D. O_2
E. There is no limiting reagent.

2. Atoms of three different elements are represented by O, Ø, and ●. Which is the limiting reagent when two OO molecules and two ØØ● molecules react to form OØ● and ØØ.

A. OO B.* ØØ● C. OØ● D. ØØ
E. There is no limiting reagent.

Figure 3. Question pair for limiting reagent.

1. What is the empirical formula of a compound if a sample of the compound contains 1.0×10^{23} Si atoms and 0.50 mol of Fe atoms?

A. FeSi B. $FeSi_2$ C. $FeSi_3$ D. Fe_2Si E.* Fe_3Si

2. Two moles of H_2 gas are known to combine with one mole of O_2 gas to form two moles of a substance called water, which we write as H_2O. Which of the following concepts is NOT associated with understanding this statement ?

A. Chemical reactions involve the breaking and rearranging of chemical bonds.
B. Chemical formulas show the ratios of atoms in a molecule.
C. The moles of H_2, O_2, and H_2O are proportionally related to each other.
D.* Chemical formulas show the spatial arrangement of atoms in a molecule.
E. The number of moles of water formed are determined by the number of moles of H_2 and O_2.

Figure 4. Question pair for empirical formulas.

1. Potassium, vanadium, and iron crystallize in a body-centered cubic unit cell. Given the lengths of the unit cell edges and the atomic weights listed below, which of these elements has the highest density (is the most dense) ?

Potassium: a = 5.250 Å AW = 39.098 Vanadium: a = 3.024 Å AW = 50.942 Iron: a = 2.861 Å AW = 55.847

A. Potassium B. Vanadium C.* Iron
D. They all have the same density.
E. Not enough information is given.

2. The drawings below are drawn to scale and illustrate the crystal structure of rubidium, niobium, and molybdenum. The atomic weights of these elements are roughly equivalent. Which of these elements has the lowest density (is the least dense)?

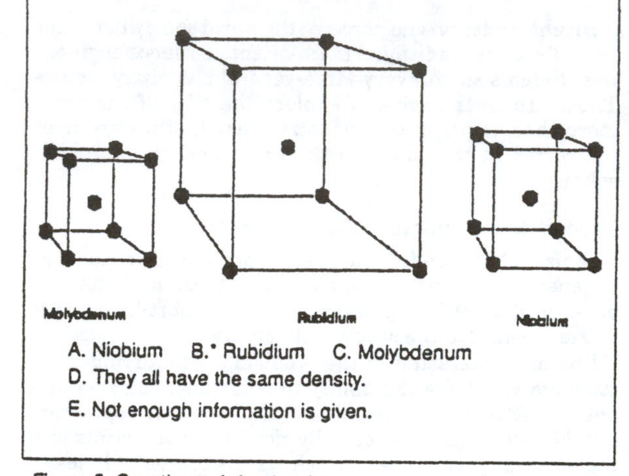

A. Niobium B.* Rubidium C. Molybdenum
D. They all have the same density.
E. Not enough information is given.

Figure 5. Question pair for density.

Figure 6 illustrates the possible categories.

Data Analysis

The responses were categorized and then frequencies were tabulated. A correct answer on a conceptual question was coded as C1; a correct answer on an algorithmic question was coded as A1. Thus, four possible combinations of responses for each pair were possible.

A1C0: algorithmic question correct; conceptual question wrong
A0C1: conceptual question correct; algorithmic question wrong
A0C0: both questions wrong
A1C1: both questions correct

Algorithmic Problem Solving	Conceptual Thinking: High	Conceptual Thinking: Low
High	Meaningful problem solving; good understanding	Many successful chemistry students
Low	Second-Tier Students who are more interested in why than how	Many unsuccessful chemistry students

Figure 6. Possible categories of students in general chemistry classes.

Table 1. The Frequencies of Response Categories by Question Pair

	students	A1[a]	C1[a]	A1C1	A1C0	A0C1	A0C0
			Gas Laws				
Remedial	167	131	32	29	102	3	33
Sci/Eng	830	715	450	397	318	53	62
Majors	56	47	23	19	28	4	5
Honors	37	35	28	27	8	1	1
			Equations				
Remedial	167	104	100	66	38	34	29
Sci/Eng	830	672	437	372	300	65	93
Majors	56	53	41	40	13	1	2
Honors	37	36	22	21	15	1	0
			Limiting Reagent				
Remedial	167	86	82	37	49	45	36
Sci/Eng	830						
Majors	56	30	26	20	10	6	20
Honors	37	33	21	19	14	2	2
			Empirical Formula				
Remedial	167						
Sci/Eng	830						
Majors	56	42	54	41	1	13	1
Honors	37	31	34	29	2	5	1
			Density				
Remedial	167	124	88	67	57	21	22
Sci/Eng	830	634	632	507	127	125	71
Majors	56	47	46	39	8	7	2
Honors	37	34	34	31	3	3	0

[a]The A's indicate algorithmic questions, and the C's indicate conceptual questions.

Table 1 gives the frequencies in each response category.

Significant Differences

By inspection, these data indicate that many students can answer an algorithmic question about a chemical idea but cannot answer a conceptual question dealing with the same topic. A chi square analysis could not be used to test these differences for significance because the same student answered both questions; the questions were not independent of each other.

Therefore, McNemar's test was used to test the significance of the differences between performance on the conceptual questions and their mathematical counterparts. The analysis tested the correlation between the number of students who answered a mathematical question of a pair correctly and those who answered a conceptual question of the same pair correctly.

The probability level was set at 0.05. Therefore, a probability less than 0.05 means that the results are statistically significant and probably not caused by chance. As can be seen from Table 2, eleven of the comparisons are statistically significant. This indicates that there was indeed a difference in performance between questions 1 and 2 in each pair.

Table 2. Significance Levels of Differential Performance on Algorithmic and Conceptual Questions for All Courses

	Remedial	Sci/Eng	Majors	Honors
Gas Laws	0.0001[a]	0.0001[a]	0.0001a	0.0391[a]
Equations	0.6406	0.0001[a]	0.0005[a]	0.0005[a]
Limiting Reagent	<0.0001[a]		0.4545	0.0042[a]
Empirical Formula			0.0018a	0.4531
Density	0.0001a	0.8503	0.7905	1.00

[a] Significant at $p \leq 0.05$.

Questions Concerning the Gas Laws

The questions on the ideal gas law were adapted from studies by Nurrenbern and Pickering (*3, 4*) and Sawrey (*5*) that involved the general chemistry classes at three other large universities. These questions were specifically included in the present study for two reasons. First, they are questions that probe an area of chemistry found in every type of introductory course. Second, the frequencies and significance levels they reported for their question pair were similar to the frequencies and significance levels for the gas law questions generated by the present study. This comparison provided a useful check on the validity of the data in the present study.

Nurrenbern and Pickering reported significant differences ($p \leq 0.05$) between conceptual and algorithmic gas law questions for all groups in their study (N = 205). In their study, 65% of the students in all categories could work the algorithmic problem, but only 35% could correctly respond to the conceptual question. This corresponds very well to the data in this study. From Table 1, a simple calculation indicates that 85% of the students (N = 1,090) could successfully answer the algorithmic gas law question (A1), but only 49% could correctly answer its conceptual counterpart (C1).

In answering the conceptual and algorithmic questions, students in this study fared slightly better than their counterparts in Nurrenbern and Pickering's study, but the basic pattern is the same. This comparison provides some evidence that the differences in performance found in this study are indeed valid and that they follow a trend noted in other large schools with general chemistry programs.

Results

Three interesting findings are noted in Table 2. First, performance on the questions for equations are not significantly different in the remedial course, although the initial hypothesis stated that differences were expected. This might be explained by the fact that the professor in the course tries to incorporate conceptual ideas in both the lectures and the exams.

Second, the fact that chemistry majors had no significant differences in their performance on limiting reagents is interesting and largely unexplained. Because every student in the course had declared chemistry as a major, perhaps these students have been more willing to construct their knowledge in terms of atoms and molecules.

Third, Table 2 shows no significant differences in performance on density questions except in the remedial course. These questions probed algorithmic and conceptual understanding of density. A reasonable speculation is that the concept of density has been presented to students many times in their high school career and that many students

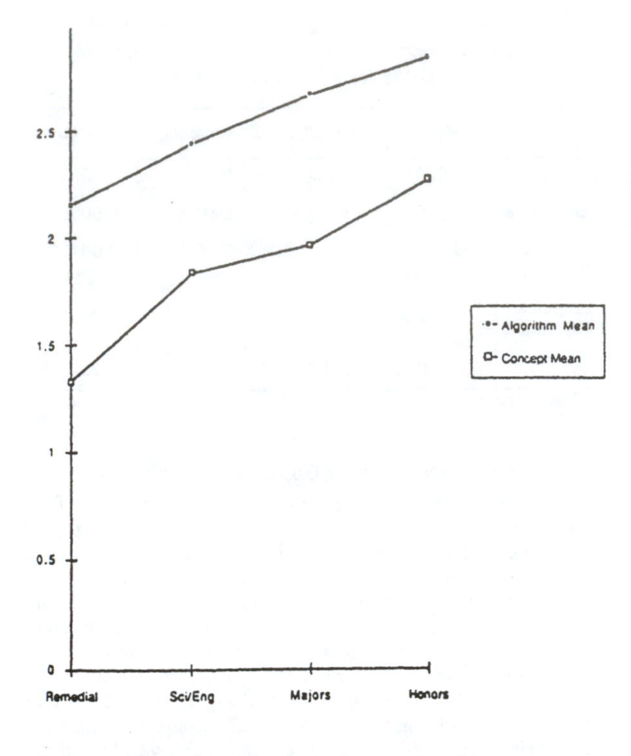

Figure 7. Algorithmic and conceptual means for all courses on gas laws, equations, and density.

had gained both algorithmic and conceptual understanding of the topic.

Comparing across Courses

The pairs dealing with gas laws, equations, and density were chosen to investigate differences across courses because only these three pairs were included on every exam. Significant performance differences were obtained across courses for gas laws ($p = 0.016$), equations (p), and density ($p = 0.006$).

Figure 4 depicts the mean score on algorithmic and conceptual questions for these three pairs. Because there were three pairs, the top score in each category was three. Figure 7 clearly shows that there were significant differences in performance within courses and across courses. These data support the initial hypothesis that there would be differences in performance across courses.

Assessment by Quadrant

This strategy of randomly placing paired questions on examinations provides a reasonably quick way to assess the understanding of general chemistry students at the algorithmic and conceptual levels. Students did seem to divide into four groups.

When all responses were counted, 49% fell in the high algorithmic/high conceptual (A1C1) quadrant of the table, and 31% were in the high algorithmic/low conceptual (A1C0) quadrant. Fewer responses (10%) fell into the low algorithmic/high conceptual (A0C1) quadrant, where it was originally thought second-tier students would reside. Also, only a few responses (10%) were categorized as low algorithmic/low conceptual (A0C0).

However, the fact that a full 31% of the responses were located in the low conceptual/high algorithmic quadrant of the table is a cause for concern. Apparently many able students were leaving their first semester of chemistry with good algorithmic problem-solving skills but weak understanding of concepts and principles.

Reviewing the Hypotheses

The honors students tended to have greater success in answering both the algorithmic and conceptual questions, which supported the initial hypothesis. However, the remedial students did not have greater success with the conceptual questions as opposed to the algorithmic questions. This finding does not support the original hypothesis. Finally, the science/engineering majors course did appear to contain a mix of algorithmic and conceptual students, which again supports the original hypothesis.

Investigating Student Preferences

This study has indicated directions for future research. Students' preferences for either conceptual thinking or algorithmic problem solving were not investigated. This might be a critical issue because Tobias (*1*) indicates that her second-tier students often could work the problems but found them boring. Her students wanted to learn more about the theories and principles of chemistry and less about the strategies used in problem solving.

The best way to investigate this issue would be to conduct individual interviews, but some information could be gleaned from an exam question asking students to indicate their preferences. These answers could then be correlated with their performance on the algorithmic vs. conceptual questions. Also, follow-up interviews could be conducted with students from each quadrant in the truth table to determine if the students had been indeed correctly identified by the test.

Conclusion

Chemistry apparently appears less and less interesting or inviting to large segments of our student population. The current emphasis on algorithmic problem solving in our general chemistry courses may have something to do with that trend. By integrating both the conceptual and algorithmic understanding of a topic in chemistry and then testing that integration, instructors might make chemistry more attractive to second-tier students.

Acknowledgment

This study was supported by funds provided by Sheila Tobias. The author gratefully acknowledges the assistance of graduate student Phemie Dandashli in analyzing the data. The author also thanks George Bodner and Bill Robinson of Purdue and Maurice Schwartz of Notre Dame for suggestions for questions.

Literature Cited

1. Tobias, S. *They're Not Dumb, They're Different: Stalking the Second Tier*; Research Corporation: Tucson, AZ, 1990.
2. Heylin, M. *Chem. and Engn. News* May 20, 1991, 29–30.
3. Nurrenbern, S.; Pickering, M. *J. Chem. Educ.* 1987, *64*, 508–510.
4. Pickering, M. *J. Chem. Educ.* 1990, *67*, 254–255.
5. Sawrey, B. *J. Chem. Educ.* 1990, *67*, 253–254.

Reflections on Leader Training

Noah Burg

In an early session of the Workshop Leader's Training course (at CCNY), held once a week for the student leaders, we did an exercise where we broke up into pairs: one of us was supposed to solve a chemistry problem and the other was supposed to listen (and possibly provide guidance if their partner got stuck). I had always fancied myself a patient person. I believed that I possessed the necessary qualities for being an effective workshop leader: I am genuinely interested in helping people learn, I am passionate about the subject matter, and I know the material. This exercise taught me a valuable lesson to the contrary, and I know that both my workshop students and I have benefited.

We were given a problem that was designed to test our knowledge of the concept of equilibrium. My partner was supposed to solve the problem and I was supposed to be the listener. She gave an answer that I knew to be incorrect. I initially attempted to direct her to the right answer without telling her, suggesting that she look at a certain aspect of her premise which was incorrect. This proved fruitless, and I then told her that she needed to redo part of her answer. This was met with a great deal of resistance. She wanted to stick with her answer and was not interested in alternatives. I could not convince her to come up with another solution, so I told her what I thought the answer should be and why. When the whole group reconvened and discussed the outcome of our exercise, my partner criticized me for being too quick with the answer; I in turn criticized her for not trying hard enough.

Upon reflection I realize that the weight of the situation would have rested almost entirely on me as a workshop leader in a real classroom situation. I have a hard time keeping my mouth shut when I know the answer and I see someone struggling with it. While I recognize how important it is not to give away answers to the students, the class proved pivotal in making me a more effective educator.

What we are trying to do is instill students with self-sufficiency, giving them the skills they need to think critically and solve problems on their own. Divulging answers only serves to reinforce their reliance on me for those answers and they tend to call me over whenever they are stuck. I also blamed myself for not encouraging my partner properly. I believe it is the job of the workshop leader to cajole reluctant students to participate, to reap the full benefits of the workshop program. To this end the workshop leader must be adept at evaluating the students on a case-by-case basis. Some will respond to the tough-love approach, while others may need to be treated gently and delicately.

The simulation in class allowed me not to make these mistakes: I was able to learn and correct for them before I led my workshop session. Based on student reactions and my own observations, the workshop session went more smoothly than I could have hoped. I was able to tell the students in the very beginning that I was not there to give answers, but to show them how to find them. This attitude initially met with resistance. Many students felt that they had tried the problems and could go no further. With the help of an initial icebreaker where the students introduced each other, I am happy to report that by the end of our initial two-hour session, students were working together quite effectively at solving the problems and not asking for my help.

Our Biology workshop program is a pilot and attendance is not mandatory, so it was encouraging to find that the workshop attendees' average performance on their first exam was above that of the class as a whole.

I believe it is a very important skill for an educator to learn how to properly motivate and captivate the interests of the students on a particular subject matter, but in all fairness it is also the students' job to meet the teacher halfway. The workshop program provides a medium where a student is encouraged to go the distance; it serves to bridge the gap between the lecturer and the students, and benefits all involved. In my experience class discussions become more meaningful because the general level of preparedness goes up, the students improve at problem solving, study groups form, motivation and self-reliance are enhanced. The workshop leaders benefit too. Their participation enables them to reinforce the knowledge of the material in their own minds, as well as derive a feeling of self-satisfaction from helping others learn.

Concept Maps in Chemistry Education*

Alberto Regis and Pier Giorgio Albertazzi
Istituto Tecnico Industriale "Q. Sella" - Via Rosselli 2, 13051 Biella (Bl) Italy

Ezio Roletto
Universita di Torino, Dipartimento di Chimica Analitica, via Pietro Giuria 5, 10125 Torino, Italy

During the first half of this century, ideas about the nature of scientific knowledge radically changed owing to the work of epistemologists and science historians such as Bachelard, Koyré, Cassirer, Popper, Kuhn, Lakatos, Feyerebend, Laudan, and Putnam. At the beginning of this century, epistemologists held an empiricist/positivist conception of science. This latter was conceived as a realistic description of the world "as it is," a body of established knowledge obtained by uncovering scientific principles (concepts, laws, and theories) "hidden" in nature. Scientific knowledge was considered the result of inductive inferences, starting with simple, unprejudiced observations, the secure base from which generalizations may be drawn, leading infallibly to conceptual explanations.

Science is at present conceived as a human activity, a "fabrication" of scientists, elaborating "models" for interpreting the empirical world and for inventing new experiments (1). According to contemporary philosophical views, scientific principles do not find their source in the facts, but they are invented by scientists to give significance to the facts. Science is not the result of inductive inference, but a hypothetical knowledge fabricated by human beings in order to understand the world and put some order in it.

These revolutionary changes in the conception of the nature of scientific knowledge (in the field of epistemology) have been accompanied by radical changes in the conception of how learning occurs (in the domain of educational psychology): the dominant view is no more the behavioral psychology but the cognitive one. Learners are actively engaged in constructing knowledge: the acquisition of new knowledge has to be firmly anchored to existing concepts, and conceptual frameworks play a key role in the acquisition, retention, and application of new conceptual knowledge and in the problem-solving exercises of the school laboratory (2). Epistemology and educational philosophy have then a common ground. Science is fabricated by scientists, in order to understand the world and to make predictions on natural and artificial phenomena, moving from the scientific principles already defined by the scientific community. Scientific learning is constructed by students starting from their "initial" conceptions of a subject matter.

More than twenty years of research on students' alternative frameworks leads to the conclusion that teachers have to take them into account if they want to help learners to acquire meaningful scientific knowledge, so proving the soundness of Ausubel's fundamental assumption of cognitive learning:

> *The most important single factor influencing learning is what the learners already know. Ascertain this and teach accordingly (3).*

To be successful in learning, students have to take possession of knowledge actively, by seeking explicit, conceptual linkages between new concepts and those they already possess. This process of elaborating personal, meaningful knowledge takes place by restructuring the already existent conceptual frameworks.

---

* *J. Chem. Ed.* **1996**, *73*, 1084-1088

The concept map (CM) is a tool, based upon the cognitive psychological theory of constructing meaning, developed by Novak and Gowin (4) as a convenient and concise representation of the learner's concept/propositional framework of a domain-specific knowledge. The concepts with their linking relationships would be "visible" in a CM as *concept labels* and *verbal connectives*, illustrating the organization of the concepts in the learner's cognitive structure. It would then be possible, at least partly, to follow the restructuring and the evolution of the cognitive structure by comparing successive CMs elaborated by the student himself at different stages of the teaching/learning process of a given topic. CMs could so reveal:

- the concepts already present in a student's mind (initial concepts);
- the conceptual linkages between the concepts (context);
- the evolution that takes place as a consequence of teaching/learning activities (conceptual change).

This is the hypothesis on which we based our use of CMs, being fully aware that CMs cannot give a complete view of the mental structure of a student.

We report here on our experience with the students (16-18 years old) enrolled in the final three years of the chemistry specialization in a technical school. CMs were used as vehicles for visualizing the students' knowledge structures and for documenting and exploring changes in these structures resulting from learning.

Training of Students

The first class sections are devoted to training the students in the concept mapping technique by introducing them to operational definitions of terms applied to CMs: concept, concept label, context, linking relationship, proposition, cross-link. Some maps are constructed on non-chemical subjects, taken from common sense knowledge. Acetate transparencies of students' constructed CMs served as models for discussion. The extent of practice needed for students to acquire proficiency in the construction of CMs depends on several factors, even psychological ones. On the average, four to six normal class sessions (45 minutes each) are required by students to learn how to construct CMs.

How to Make Concept Labels Available to Students

Three different ways of assigning concept labels are used, leaving the students the choice of the linking relationships.

Known Terms CM

At the first stage, students are invited to produce what we call a *known terms CM*. In this case:

- a fixed number of concept labels are assigned;
- each student constructs his own map by using only the terms given, choosing the linking relationships and the concept structure he considers most suitable.

For example, third-year students who had already followed a first course on general chemistry were asked to develop a CM on the topic "oxidation-reduction" using the following concept labels: atom, battery, (electrical) current, electrical conductor, electrolysis, electron, ion, nucleus, oxidation, reduction. Since the objective was to point out the eventual changes in the cognitive structure following teaching of the topic, the same terms were given before and after teaching. The maps were drawn up individually during a normal class session. During the drawing up of the post-instruction map, students did not have their pre-instruction CM at their disposal.

By comparing the two maps drawn up by each student, changes in the structure of the maps were found in more than three quarters of the students, even if not all the changes were improvements. It is reasonable to think that changes in the maps correspond to similar changes in the conceptual structure of the students due to learning. In Figure 1 two successive maps, produced by the same student, are shown. The concept label electron, which is linked only to current and atom labels in the first map, has four other linking relationships in the second one.

The changes in the links between the labels battery, current and electrolysis are also interesting. In the second map, it is the battery that generates the electrical current and not electrolysis. In one class, after teaching, 19 of 24 CMs linked electrolysis with battery wrongly; the linking relationship explained that battery worked by means of electrolysis. This was interesting since:

- electrolysis had never been mentioned in our lessons, and we had deliberately inserted this concept label, which is dealt with in other courses (physics, for example);
- the students had come from different classes with different chemistry teachers.

The very high frequency of the wrong connection suggested that there must have been some event that had influenced all students in the class. During the discussion of the maps, we found that the error was the consequence of the teaching in another course: the teacher had certainly been successful!

Analysis of these "noted terms CMs" gave us the possibility of recognizing a first type of cognitive event, which we called "*Cognitive Event fix*" *(CEfix)*. A concept that is incorrectly inserted in a student's conceptual structure is no longer accepted in the same position after learning (Figure 2). The destabilization of the mental structure due to meaningful learning has a favorable outcome when the concept in question is stabilized correctly. The connected concepts are partially reordered so that the conceptual framework is restructured.

This type of map allows one to follow the evolution of the cognitive structure after the teaching/learning activity. Moreover, since the number of concept labels is fixed, this kind of map is particularly useful for identifying recurrent alternative conceptions held by students.

Guided Choice Terms CM

A second way of assigning concept labels is to give the students a number of terms greater than what is requested to construct the map. The criteria to follow is:

- a fixed number of concept labels related to the knowledge being investigated are assigned - e.g., 20;
- each student must choose only a fixed part of these labels—e.g., 10 - to construct his own map using the linking relationships and structure that he considers most suitable.

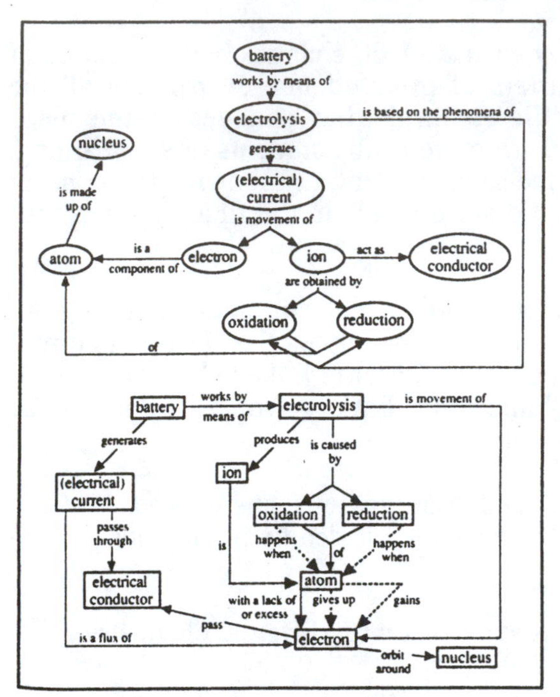

Figure 1. Oxidation and reduction: two successive maps produced by the same student.

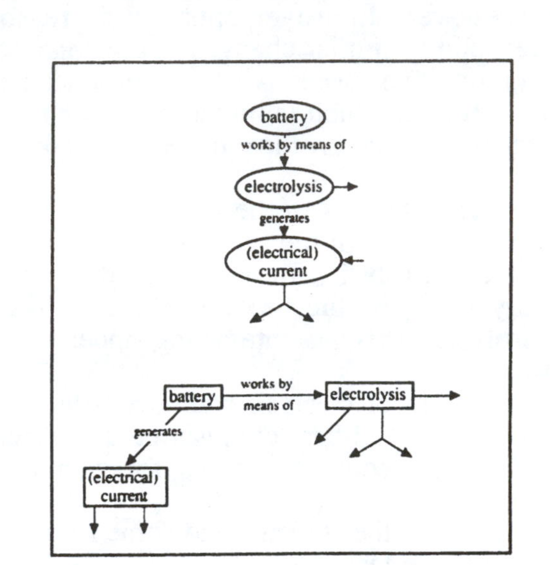

Figure 2. Cognitive Event fix (CEfix). In this case, a concept which is incorrectly inserted in a student's conceptual structure is no longer accepted in the same position.

For example, third-year students were invited to construct a *guided choice CM* on atomic structure during a normal class session. Before teaching the topic, the following twenty terms were supplied to assist the subjects with the task: atom, Aufbau principle, chemical bond, corpuscle, electron, energy level, exclusion principle, Hund's rule, hybridization, ionization energy, nucleus, orbital, orbit, quantum numbers, shell, spatial geometry, spin, uncertainty principle, wave, and wave function. The students were then asked to choose ten terms and develop a CM. This procedure was repeated at the end of the teaching/learning activities, leaving the students the possibility of using ten terms they considered most suitable—not necessarily the same terms used for the first map, which was not at their disposal.

Figure 3 summarizes the number of students who used the different concept labels in the CMs. The darker bars refer to the choices made for the first map (before teaching) and the lighter ones concern the choices made for the second map (after teaching).

Analysis of the maps shows that before teaching, seven students out of ten chose the concept label "orbital," but no one associated it with the label "wave function." Four of these seven use the labels "orbit" or "shell" to structure their knowledge. The concept "orbital" does not seem to be an alternative one for "orbit," but rather complementary. After teaching, the label "orbital" is used by ten out of twelve students; three of these ten still associate this concept with the label "orbit," while the label "wave function" is still not used by any student.

Figure 4 shows two maps that were constructed by the same student before and after teaching. They are quite different.

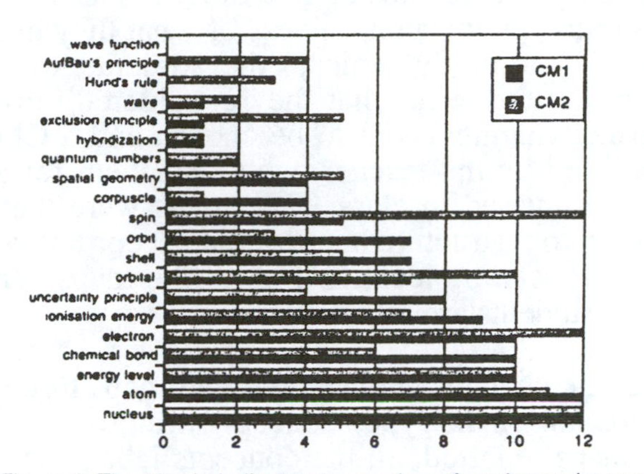

Figure 3. The *X*-axis indicates the number of students who used the various concept labels that are indicated on the *Y*-axis. The darker bars show the original choice and the lighter ones represent the choice made at the end of the unit.

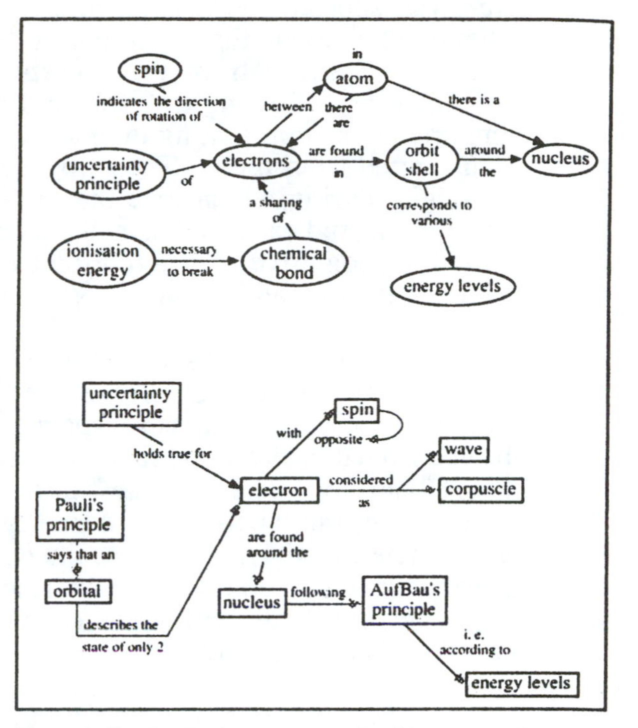

Figure 4. Atomic structure: an example of two successive maps constructed by the same student.

We can identify two types of cognitive events in these maps. The first, which we call *"Cognitive Event in" (CEin)*, occurs when a concept label, in this case "orbital," enters the conceptual framework (Figure 5). The second, which we call *"Cognitive Event out" (CEout)*, occurs when a concept label, in this case "Orbit," of the first map is excluded from the second one (Figure 6). As can be seen, the concept "orbit" is not simply replaced by the concept "orbital," since the latter occupies a different position with completely different links. It can reasonably be assumed that the student's cognitive structure is undergoing a conceptual reconstruction.

Concept Stimulus CM

In the preceding approaches, the teacher gives the students the concept labels referring to a specific domain of chemical knowledge. But it is also possible to ask the students to identify the most suitable or important concept labels and to construct what we call a *Concept stimulus CM*. In this case:

- only one concept label (stimulus) is assigned;
- the number of labels the students can add is fixed;

- each student elaborates his own map using the concept labels, the relationships, and the structure that he considers most suitable.

Fifth-year (final-year) students were invited to construct this kind of map on a very vast subject, thermodynamics. The aim of the teacher was to give the students, after studying this topic, the possibility of demonstrating the meaningfulness of what they had learned. Starting from the concept stimulus "thermodynamics," the students were assigned the following task for the first map: the ten terms that they considered most important for representing the basics of thermodynamics were to be chosen and a CM constructed from them. The maps were collected by the teacher, who kept them for a fortnight, during which time the subject was discussed in class. The maps were then given back, and each student had the opportunity to restructure his own map accordingly. The three cognitive events *(CEfix, CEin, CEout)* can be identified in all the maps. In Figure 7 the first and second maps of one of the students are reported.

Initially, it seems very difficult to link the changes between the two maps to the cognitive events cited before. Since the maps are of the type "concept stimulus," any term considered suitable is accepted. In the maps reported, all the concepts labels have been changed between the first and second maps. But, have they really? The labels "1st law," "2nd law," and "3rd law" in the first map have been changed into "laws" in the second one; the labels "ΔU," ΔH," "ΔG," "ΔS" in the first map are summarized in the term "state functions" in the second one. This can also be seen for the labels "chemical thermodynamics" and "chemical phenomena" in the second map, which derive from the terms "affinity" (understood as chemical affinity) and "spontaneity" (spontaneity of a reaction) in the first map.

The vastness of the topic caused some difficulty for the students, who resorted to more general concepts only in the second map, thus summarizing the large variety of terms used in the domain of thermodynamics. It must be remembered that the way of giving the students the concept labels directly influences the types of cognitive events to be found in the CMs they develop. "Noted terms" maps never show "in" or "out" cognitive events, while "guided choice terms" or "concept stimulus" maps normally show all three events. These obviously do not represent the complexity of a human's mental processes; they are extreme cases, which should be looked for in students' CMs, as they are signs of meaningful learning.

Moreover, a concept map is an idiosyncratic representation of a domain specific knowledge. Consequently, the maps reproduced are not representative or typical of our students. A CM strictly reflects the conceptual organization of the single student who has produced it, giving evidence to a specific level of conceptual understanding.

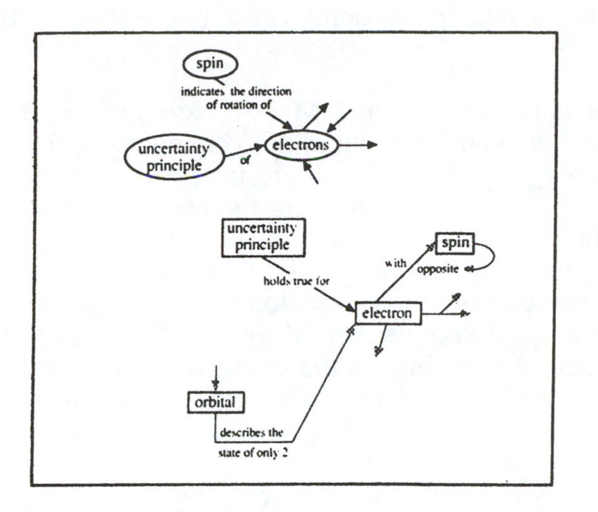

Figure 5. Cognitive Event in (CEin), occurs when a concept label, in this case "orbital", enters the conceptual framework.

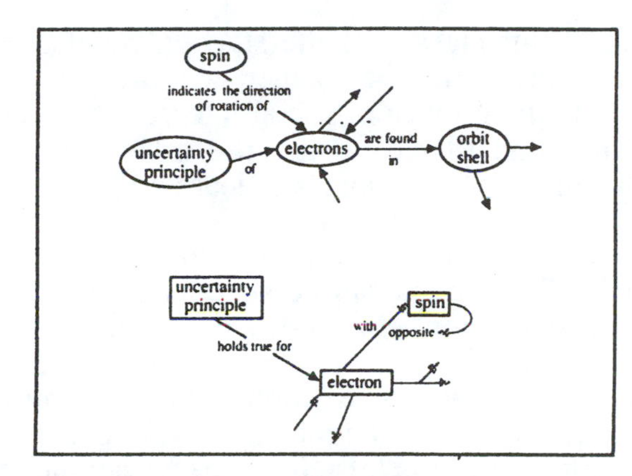

Figure 6. Cognitive Event out (CEout) occurs when a concept label, in this case "orbit", of the first map is excluded from the second one.

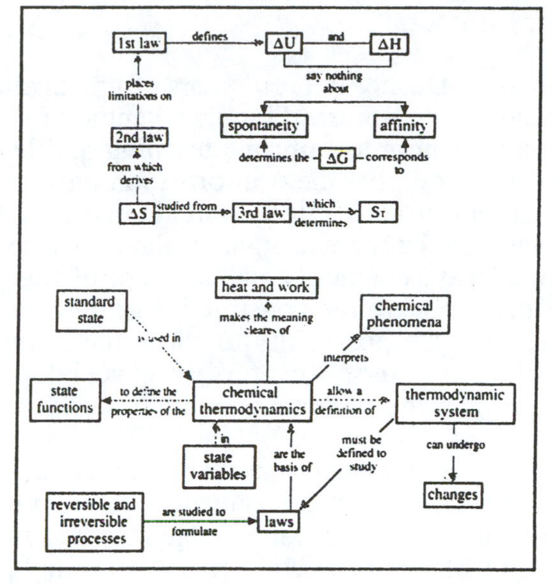

Figure 7. Thermodynamics: an example of two successive maps constructed by the same student.

Conclusions

It is now four years since we introduced concept maps into our chemistry courses, as participants in an action-research project concerning the improvement of chemical education in secondary schools. At the beginning, we tried to use them as assessment devices: scores were assigned to post-instruction maps for the number and correctness of the relationships portrayed, for the levels of hierarchy, and for cross-linking. We also tried to assign scores for the convergence of the students' maps to teacher-constructed maps. But we soon had to recognize that students' CMs are highly idiosyncratic representations of a domain-specific knowledge, and the interindividual differences displayed among them were far more striking than the similarities. Taking into account the critical opinions on the scoring of concept maps (5), this observation led us to shift emphasis and focus on changes in content and organization of CMs over time, and on helping the students to become aware of and criticize their own frames and those of the others. Such use is consistent with a constructivist theory of learning, which is the foundation of our teaching strategy, viewing knowledge as being actively constructed by the learner.

During the four years of practical experience in using CMs, we have grown more and more impressed by the potential of this metacognitive tool to help chemistry teachers and learners to improve teaching and learning. Concept maps are useful for teachers, since they give them information on what students know, showing the concepts already present in their minds (initial concepts), how they are related to one another (context), and how learners reorganize their cognitive structure after a specific teaching activity. In this way, teachers can be aware of the presence of misconceptions that are potential "obstacles" to the construction of new, meaningful knowledge. Moreover, CMs give the teacher the possibility to check the influence of teaching on the cognitive structures of students, since this tool is especially valuable in documenting and exploring the restructuring of conceptual frameworks.

But concept maps are also judged a very useful metacognitive tool by students. In fact, many of those trained to develop CMs in chemistry have spontaneously adopted them to represent knowledge in other disciplines, such as history and Italian, claiming that CMs are powerful helps in meaningful learning of new subject matter.

Concept maps are also currently used in our classes as the starting points for discussions on chemistry topics involving the students and the teacher, who acts as the chairman. Von Glaserfeld (6) has emphasized the importance of social interaction in the construction of new meanings, and we have found that by using the CMs we can effectively act both on the interpersonal and the intrapersonal aspects of learning.

As Novak says "concept map...is no 'magic bullet,' no 'quick fix' for classrooms where rote learning predominates" (7) and use of the maps can be successful only by adopting a constructivist approach to chemical education. Many teachers object that turning to a constructivist approach and using concept maps is very time-demanding, and their objection is both true and false. It is true if we limit our attention to the beginning of the course, when students have to become acquainted with the idea that they construct their own knowledge and learn to develop concept maps. It is false if we go further along the course, since the meaningful learning of the first fundaments of chemistry (the particulate model and the conceptions of pure substance and chemical reaction) gives the students powerful instruments to construct further meaningful knowledge more easily and more quickly than in traditional teaching.

Acknowledgments

Work carried out with the financial contributions of MURST (Ministero dell'Università e della Ricerca Scientifica e Technologica. Fondi 40% - Progetto: Insegnamento e Apprendi-mento della Chimica) and with the support of IRRSAE Piemonte.

Literature Cited

1. Chalmers, A. F. *Science and its Fabrication*; Open University: Buckingham, 1990.

2. Hodson, D. *Sci. Educ.* **1988**, *72(1)*, 19-40.

3. Ausuel, D. P. *Educational Psychology: A Cognitive View*; Holt, Rinehart and Winston: New York, 1968.

4. Novak, J. D.; Gowin, D. B. *Learning How to Learn*; Cambridge University Press, 1984.

5. Stuart, H. A. *Eur. J. Sci. Educ.* **1985**, *7*, 73-81.

6. Von Glaserfeld, E. *Synthese* **1989**, *80*, 121-140.

7. Novak, J. D. *J. Res. Sci. Teach.* **1990**, *27*, 937-949.

Gender and Silence
Implications of Women's Ways of Knowing*

Joan V. Gallos†

Whether teachers quote research or exchange anecdotes about teaching life, few deny the significance of gender in the college classroom. The all-too-frequently heard claims that males dominate class discussions parallel end-of-term laments about women students who turn in brilliant written work yet were silent in class discussions. How do we explain this? What are the connections between gender and silence for many women in the classroom? What are the implications for college teaching? Part 1 of this article examines women's experiences and perspectives toward learning and why the university classroom is usually more comfortable for men than for women. Part 2 explores the implications for effective teaching and for creating equal learning environments for men and women.

Part 1: Women's Learning Experiences and Perspectives

Women's Accounts, Powerful Clues

How do your teaching strategies and class designs reflect the unique learning needs of women? What are those unique needs? For most professors, such questions are hard to answer. Many are unaware of growing research on the effects of gender in the college classroom. Even those who know are often unsure of what to do about it.

Women's accounts of their experiences in adult learning provide powerful insights into their distinctive needs. Listen to one woman's story—let's call her Barbara.

> I clam up in classroom situations. I have been taking women's studies courses nights during the past year and a half for the purpose of trying to understand and work on this. The first course I took, I concentrated on just talking and spoke out about two or three times. It was so hard. The next term, I spoke more often and worked on trying to sound coherent.

Barbara is in her mid-thirties. In contrast to the silent, anxious person described above, Barbara is actually a part-time, postdoctoral student at a major university and a successful scientist and administrator at a prestigious medical school. Moreover, she is an honors graduate from a major women's college and the recipient of the top fellowship in a competitive Ph.D. program at another highly respected institution. Barbara acknowledged her fears at a "Gender and Management" workshop at her medical school.

Barbara is an effective administrator, supervisor, and researcher. She spoke eloquently throughout the workshop, discussing subtleties in the cases, examining her beliefs and behaviors, generalizing learning from personal experience, responding supportively to others, moving discussions to deeper levels and introducing relevant topics. She was confident, articulate, and unquestionably very bright. On the surface, Barbara was an instructor's dream. How could this be the same person who spoke of difficulties saying *anything* in a classroom setting? I asked Barbara. Her answer was

---

* *College Teaching* **1995**, *43*, 101-105.

† Joan V. Gallos is an Associate Professor in the Division of Urban Leadership in the School of Education at the University of Missouri-Kansas City.

quick and direct. "Oh, but this is just with women. I would never be able to do this with men."

The connections between gender and silence are powerful for Barbara, and she is not alone. Inspired by Barbara's candor, other women at the session poured out similar concerns. Many spoke publicly for the first time, surprised that others who seemed "so together" shared their hidden fears. Their words were no surprise to me. In fact, they were echoes of many other voices in my work with other all-female groups.

The Discovery Women: Powerful Themes

One poignant example was a course I taught in the Radcliffe College Discovery program for under-employed women.[1] I opened the course by asking the women to discuss how they felt. Three powerful themes dominated their responses—great fear, self-doubt, and novelty. The women were terrified that they wouldn't be able to understand, wouldn't know what to do, and would demonstrate that they did not belong. The doubted themselves and were sure that they would have nothing to say. They also experienced a sense of novelty—that this would be like nothing they had ever done before. I named the themes, filed them in the back of my mind, and moved on to introductions.

I asked the women to introduce themselves and say something about their goals for the course. I wrote on the board: Who am I? My goals? My expectations?

I had not anticipated how powerfully those questions tapped into the fear described earlier. Despite differences in age, race, ethnicity, education, marital status, and socioeconomic class, woman after woman acknowledged that she did "not know" who she was. Some spoke of hiding behind the title of mother, wife, daughter, or student only to be left empty when role demands changed. Many shared bitter tales of powerful others with strong expectations for what the women should—or should not be—and of maintaining fluid identities as a survival tactic—a way to avoid the pain and punishment from becoming one's own person. The anguish behind the women's stories revealed the toll. With courage, they had come forward to gain new skills and, as they now told me, to claim themselves at last.

After the first class, several women marveled that all this was different than expected. One women stated with surprise that "what I already know and feel would be relevant to what I would learn at a place like Harvard." They were surprised that they already knew something useful, that their experiences counted and could be foundations, not embarrassments, for more learning.

What they knew mattered. That explained the novelty theme. These students had years of experiences in schools and classrooms. Why would they think they knew nothing relevant? Why is such self-doubt and alienation, not a Discovery anomaly, but many women's frequent companion in academic settings?

Self-Doubt

Women harbor more self-doubt and questions about capabilities and intellectual competence than men do. A major, multi-year study of Harvard undergraduates (Light 1990) supports the claim. Despite educational achievement, success, or satisfaction, when the Harvard women experienced failure, they were quick to doubt themselves.

They attributed their problems to self-limitations and personal inadequacies. In sharp contrast, their male peers put the blame on others or on circumstances.

The Harvard study is powerful when we consider that the gender findings were unanticipated. The researchers had not set out to study gender differences. They were studying women and men who grew up in an era marked by liberation, change, progress, and women's advancement. The Harvard women excelled in academic environments. They were successful in making the grade, had little experience of academic failure, and had chosen to study at one of America's elite institutions. Yet these highly selected women still carried the same doubts and questions about their intellectual capabilities as the Discovery women, as their sisters in the six other educational institutions that Belenky and others (1986) interviewed, as the women in the Belenky et al. study who never set foot in a college classroom, and as successful women managers like Barbara and her colleagues. Why was this markedly diverse group of women united only by their fears and self-doubt?

Social forces combine to make women doubt their academic abilities. We live in a society where it is still easier for women to gain approval and attention for their bodies and physical attributes than for the quality of their minds. Cultural and historical definitions of femininity reinforce this message. Lists of traditional feminine stereotypes read like recipes for Western anti-intellectualism (Gallos 1982). We send modern women mixed messages: ignore the stereotypes but remain feminine. The Discovery women reported feeling punished for doing either. Such punishment is crazy-making. Self-doubt is more logical and safe alternative.

Early school experiences reflect society's mixed messages about women's fitness for learning. In a word, schools shortchange girls. A growing body of research points to this powerful reality (American Association of University Women 1990; American Association of University Women and Wellesley Center for Research on Women 1992; Wellesley College Center for Research on Women 1992; Sadker and Sadker 1994). From pre-school through high school, most girls face teacher behavior, attitudes, and pedagogy that favor boys. It does not take long for girls, who like young male peers are developmentally predisposed to see external events as personally caused, to translate their marginal status into questions of *what's wrong with me?* Most girls emerge from early school experiences with half the confidence and self-esteem of boys (American Association of University Women 1990). Most women carry this harmful legacy into the adult classroom, compounded by years of cultural and personal experiences that support their public silence (Cameron 1990; Gilligan 1997, 1980, 1982; Moulton 1989; Rosaldo and Lamphere 1974; Thorne and Henley 1983).

Alienation in Academic Settings

The male-based focus of educational systems, structures, and pedagogies contributes to women's alienation. The Harvard study (Light 1990), for example, points to gender differences in the ways that men and women approach and prepare for the classroom and relate to faculty members. Yet most modern classrooms are structured to match better the needs and expectations of men than women (Chira 1992). What happens to those whose needs stand in contrast to systems and institutions that present their ways as *the* road to learning? They feel alienated, like strangers in a supposedly familiar land.

But Belenky et al. (1986) and others (e.g., Berman 1989; Gallos 1989; K. Gergen 1988; M. Gergen 1988; Harding 1991; Hubbard 1988; Keller 1989; Noddings 1984; Rosaldo and Lamphere 1974) see the problem as deeper. Societal conceptions of knowledge, learning, and individual development are androcentric. Men have historically been the "fact-makers," in Hubbard's terms (1988). They have designed and conducted the research, served as research subjects, proposed the theories, written the histories, defined the procedures for science and instruction, established standards, controlled access to institutions, and set the public policies.

Women have been asked to learn the experience of men and accept it as representative of all human experience. When women cannot match this (masculine) knowledge to their own lives or see it as relevant, the—not the facts, theories, and curricula—have been termed deficient (e.g., Kohlberg 1981; Vaillant 1977).

The implications for women's learning are clear. As one of Barbara's colleagues explains:

> There is a self-consciousness for women speaking about their experiences [when they are] in real classrooms. In mixed groups, it's hard to say *I'm thinking about this differently.* That recognition can lead to tension, possibly conflict. I just keep things to myself and avoid it all.

High standards, strong expectations, and demands for teacher-defined "quality work," for example, seem to be gender-neutral. They propelled the men in Perry's (1968) classic study of the college years into higher stages of development. But strong standards, demands, and expectations are double-edged for women. They are essential parts of the guidance necessary for learning from a knowledgeable and experienced teacher. They are also impediments to independent thinking when most women's efforts to learn become mixed with efforts to please. If this is so, what does work for women?

A community of support and confirmation is essential for most women's intellectual growth. Major developmental studies of women (e.g., Baruch, Barnett, and Rivers 1983; Gilligan 1977, 1980, 1982; Lyons 1988) all point to the importance of relationships and an ethic of caring, not as a substitute for accomplishment and rational discourse, but as an essential complement.

Learning in community, however, stands in sharp contrast to the debates, devil's advocacy, confrontations, and individual testing viewed as essential components of a "stimulating" educational environment. Women in the Belenky et al. (1986) research, for example, found these activities induced doubt and were perceived as personal put-downs that fed their deepest fears. They spoke of learning best from a pattern of "confirmation-evocation-more confirmation": when instructors conveyed, "That's good thinking, now think more." Like the Discovery women who relished the public acknowledgment that what they knew mattered, these women wanted and needed a supportive environment to learn. Women in the Harvard study (Light 1990) made similar requests.

Part 2: Implications for College Teaching

Multiple implications for good teaching flow from an understanding of most women's preferred ways to know. What can individual teachers do?

Steps Teachers Can Take

One place to begin is to rethink readings and course materials. Professors must examine the messages about gender in readings and assignments and look candidly at what is said—and what is not said—about men and women. Are successful women, for example, well represented in cases and text illustrations? Are they presented as exceptions, as opportunities to explore only "women's issues," or as illustrations for more generic subject matter?

Professors must recognize the subtle ways in which their course materials reinforce maleness as the norm. The simple use in readings, for example, of the terms *author, scientist,* or *leader* when referring to men, in contrast to *woman author, woman scientist,* and *woman leader* when discussing women in similar positions, may seem innocent. But such terms reinforce social perceptions of achievement and success as a largely male world and send powerful messages to women about their second class status.

Expand Examples and Discussions

We teachers need to expand our examples and discussions beyond the traditional world of men's professional work, histories, sports, and hobbies. We must make our classroom illustrations alive and relevant to all students by drawing on the full range of life experiences that women *and* men bring to the classroom. We are often too limited in defining what is appropriate for discussion and too quick to assume that personal topics like family management, parent-child issues, relationships, and self-image are inappropriate.

By drawing our examples and discussions from the day-to-day lives of our students, however, we offer opportunities for all to see their varied life experiences as relevant to our histories, theories, and modes of analysis. But perhaps more important, we need to see and acknowledge their relevance and importance first. In doing that, we move beyond the implicit message to women that their unique history, training, and concerns are less important than those of men.

Redefine the Teacher's Role as a Learning Model

We must re-examine the ways in which we interpret the professorial role. We need to demystify it; moving away from teaching styles, such as the traditional lecture, that convey the image of an all-powerful and infallible expert and toward teaching strategies that model how to learn. Demonstrations, experiments, simulations, and student/professor role plays enable us to learn with and in front of students.

When we speak openly about theory building and other creative enterprises, we teach students that everyone struggles in searching for and articulating truth. When we demonstrate how we might think through a problem or case, rather than presenting finished arguments and traditional answers, we remind students that theories and models are created by people working together. Such reminders are essential for women, who have been told for generations that they are intuitive (rather than rational), and who may therefore infer that the academy's ways of learning are just too difficult.

Insight into many women's usual ways of knowing shows us how easily efforts to learn can be thwarted by efforts to please. Instructors, therefore, must encourage students to design projects to foster their own learning. During part of the course, at least, we need to offer opportunities for students to set their own standards on assignments and activities, generate criteria for evaluation, and revise strategies for assessing success. Such self-generated efforts make assignments and academic progress mean more than meeting the expectations of powerful others—a condition that has always been a reality for women (Miller 1976; Hinckley 1980; Josefowitz 1980).

Use Experiential Activities and Small Groups

Preferred ways of knowing for many women fit well with experiential learning, which asks students to enmesh themselves in a class activity, reflect on it, generalize about implications, and incorporate the new insights into future action (Kolb 1974).

Experiential learning conveys confidence in students' abilities to participate actively in their own learning. Rather than asking them to ignore their personal experiences, we should challenge our students to find their meaning and use them to understand how the world works. Learning is grounded in experience, and experience is informed by learning. This is the two-way street that many women often miss in lecture classes, competitive discussions of case scenarios, and other more traditional, hierarchical teaching.

The Harvard study (Light 1990) confirmed that men and women learned best when they worked in small groups. What is surprising is the finding that women are *less* likely than men to initiate and join small learning groups but, when encouraged to do so, they profited as much as men. Professors must recognize women's tendency to "go it alone" and create opportunities for men and women to build peer relationships and small discussion groups both in and out of the classroom.

Increased Opportunities for Encouragement and Support

Many women need a supportive classroom to learn. I am struck, for example, when watching good teaching how often instructors ignore or neglect good answers, focusing instead on those that are off-target or incomplete. We need to articulate more often, more forcefully, and more publicly the praise and acceptance formerly conveyed with a smile or a nod. We need to reinforce participation, returning by name to specific student comments to illustrate key points or central ideas.

It is easy in teaching to focus on all that still needs to be done and to lose sight of accomplishment. Instead, we must identify successes in the classroom. We can, for example, name the ways in which a strong individual or group comment shows improvement over a past question or statement of confusion. I begin each class with a short list of our progress from the last class and my sense of how this has prepared us for present work. As the term progresses, I ask students to do this progress report.

The Pygmalion effect is a powerful one. I work with students to identify some insight in whatever they say and pull out a kernel of "truth." This is critical for those whose anxieties and self-doubt interfere with class participation. I also watch for signs that students may question whether they have the "right stuff" for learning and remind them respectfully and forcefully of the strengths they bring.

Finally, we must separate feedback from criticism. Quality feedback is not evaluative; it supports and encourages learning. It describes behavior, like holding up a mirror so that one can see what others experience and understand. Good feedback provides essential, timely, and on-going information.

The literature on gender issues notes the subtle ways in which teachers treat men and women differently. Some typical patterns are making more eye contact with male students, calling less often on women students, allowing men to call out answers while women raise their hands, and offering more precise feedback or praise to men than women. Because many gender-linked teaching behaviors are subtle and non-conscious on the part of the teacher, instructors may want to videotape themselves in the classroom and work with trusted colleagues to process their observations. Sadker and Sadker (1994) offer tips about where to begin.

Build Communities of Learning

It is a challenge to build the classroom communities that many women see as essential to learning. Many students, who look to the instructor for knowledge and guidance, cannot see the importance of learning with and from peers. We need to encourage and form collaborative learning groups, devote time to team-building efforts, work with students to develop skills in effective group work, and foster a positive interdependence among students (Johnson and Johnson 1993).

It is easier for students to appreciate community-based learning when they recognize how individual efforts contribute to the whole. I see part of my teaching role as helping the class build a shared history of their learning progress with the contributions of various members clearly evident. When students recognize the benefits, they become more self-motivated and less dependent on the teacher, and they learn to respect the power of community.

This article has explored the implications for college teaching of many women's ways of knowing. It is my conviction that we need to identify the "man-made" educational trappings that encourage women's silence and that they have long been promoted as the only road to learning. We have examined what we know about most women's experiences in the classroom and implicitly have questioned how much we know about men's. Clearly there is more work to be done; however, by acknowledging that gender plays a powerful role in the classroom and by responding to the unique learning needs that women *and* men bring, we move closer to creating equitable learning environments for all our students.

Note

1. For a more detailed exploration of the Discovery case, see: J. Gallos, "Women's Experiences and Ways of Knowing: Implications for Teaching and Learning in the Organizational Behavior Classroom," *Journal of Management Education 17:1* (February 1993):7-26.

References

American Association of University Women. 1990. *Shortchanging girls, shortchanging America*. Washington, DC: AAUW Educational Foundation.

American Association of University Women and the Wellesley Center for Research on Women. 1992. *How schools shortchange girls*. Washington, DC: AAUW Educational Foundation.

Baruch, G., R. Barnett and C. Riers. 1983. *Life prints: New patterns of love and work for today's women*. New York: New American Library.

Belenky, M., B. Clinchy, N. Goldberger, J. Tarule. 1986. *Women's ways of knowing: The development of self, voice, and mind.* New York: Basic Books.

Berman, R. 1989. From Aristotle's dualism to materialist dialectic. In *Gender/body/knowledge: Feminist reconstructions of being and knowing.* ed. A. Jaggar and S. Bordo. New Brunswick, NJ: Rutgers University Press, pp. 224-255.

Cameron, D., ed. 1990. *The feminist critique of language: A reader.* New York: Routledge.

Chira, S. 1992. An Ohio college says women learn differently, so it teaches that way. *New York Times.* 13 May.

_____. 1989. Exploring women's development: Implications for career theory, practice, and research. In *Handbook of career theory: Perspectives and prospects for understanding and managing work experiences.* ed. M. Arthur, D. Hall, and B. Lawrence. Cambridge: Cambridge University Press, pp. 110-132.

_____. 1993. Women's experiences and ways of knowing: Implications for teaching and learning in the organizational behavior classroom. *Journal of Management Education 17(1)*:7-26.

Gergen, K. 1988. Feminist critique of science and the challenge of social epistemology. In *Feminist thought and the structure of knowledge.* ed. M. Gergen. New York: New York University Press, pp. 27-48.

Gergen, M. 1988. Toward a feminist metatheory and methodology in the social sciences. In *Feminist thought and the structure of knowledge.* ed. M. Bergen. New York: New York University Press, pp. 87-104.

Gilligan, C. 1977. In a different voice: Women's conception of self and of morality. *Harvard Educational Review 47* (4 November):481-517.

_____. 1980. Restoring the missing text of women's development in life cycle theories. In *Women's lives: New theory, research, and policy.* ed. D. McGuigan. Ann Arbor: University of Michigan Center for Continuing Education of Women, pp. 17-33.

Harding, S. 1991. *Whose science? Whose knowledge? Thinking from women's lives.* Ithaca, NY: Cornell University Press.

Hinckley, S. 1980. The one up/one-down model. In *Paths to power: A woman's guide from first job to top executive.* ed. N. Josefowitz. Reading, MA.: Addison-Wesley.

Hubbard, R. 1988. *Some thoughts about the masculinity of the natural sciences.* In *Feminist thought and the structure of knowledge.* ed. M. Gergen. New York: New York University Press, pp. 1-15.

Johnson, D. W. and T. R. Johnson. 1993. Structuring groups for cooperative learning. In *Mastering managerial education: Innovations in teaching effectiveness.* ed. C. Vance. Newbury Park, CA: Sage, pp. 187-198.

Josefowitz, N. 1980. *Paths to power: A woman's guide from first job to top executive*. Reading, MA: Addison-Wesley.

Keller, E. F. 1989. Feminism and science. In *Women, knowledge and reality: Explorations in feminist philosophy*. ed. A. Garry and M. Pearsall. Boston: Unwin Hyman, pp. 175-188.

Kohlberg, L. 1981. *The philosophy of moral development*. San Francisco: Harper and Row.

Kolb, D. 1974. On management and the learning process. In *Organizational psychology: A book of readings*. 2nd ed. ed. D. Kolb, I. Rubin, and J. McIntyre. Englewood Cliffs, NJ: Prentice-Hall, pp. 27-42.

Light, R. 190. Explorations with students and faculty about teaching, learning, and student life: The first report. Cambridge, MA: The Harvard University Assessment Seminars.

Lyons, N. 1988. Two perspectives: On self, relationships, and morality. In *Mapping the moral domain: A contribution of women's thinking to psychological theory and education*. ed. C. Gilligan, J. Ward, and J. M. Taylor. Cambridge, MA: Harvard University Press, pp. 21-48.

Miller, J. B. 1976. *Toward a new psychology of women*. Boston: Beacon Press.

Moulton, J. 1989. A paradigm of philosophy: The adversary model. In *Women, knowledge, and reality: Explorations in feminist philosophy*. ed. A. Garry and M. Pearsall. Boston: Unwin Hyman, pp. 5-20.

Noddings, N. 1984. *Caring: A feminine approach to ethics and moral education*. Berkeley: University of California Press.

Perry, W. 1968. *Forms of intellectual and ethical development in the college years*. New York: Holt, Reinhart and Winston.

Rosaldo, M., and L. Lamphere. 1974. Introduction in *Women, culture, and society*. ed. M. Rosaldo and L. Lamphere. Palo Alto: Stanford University Press, pp. 1-16.

Sadker, M. and D. Sadker. 1994. *Failing at fairness: How American schools cheat girls*. New York: Scribner.

Thorne, B. and N. Henley, eds. 1983. *Language, gender, and society*. Rowley, MA: Newbury House.

Vaillant, G. 1977. *Adaptation to life*. Boston: Little, Brown.

Wellesley Center for Research on Women. 1992. *Girls in schools: A bibliography of research on girls in U.S. public schools—kindergarten through grade 12*. Wellesley, MA.

The Students' Experience*

William G. Perry

The Experience (1)

In a child's first awareness of his world, what are the "terms," meaningful to him, through which he interprets his experience? It seems probable that they are the basic dual terms of his sense of well-being and dis-ease: comfort and irritation, desire and aversion, light and dark, safe and frightening. These are the terms from which he coordinates his first "purposeful" actions in his own behalf: physical movement of toward and away, social communication of appeal and rejection. Most importantly for our concerns, the child will use the simple either-orness of good and bad, permitted and not permitted, as a foundation for his first realization of himself among the people of his moral world. His imperative concern must be to know if he is to be patted or punished, praised or scolded. Is he good or is he bad?

This is not to say that in the neutrality of the play-pen children do not reveal the early development of a more complicated awareness. A child soon discovers for his own purposes that two quite different sights are nonetheless the same toy viewed from different sides. Jean Piaget has traced how the child expands this discovery into the realization of later years that the same thing can look differently to different people, a structure of thought central to our concerns.

This development, however, seems to be delayed in the sphere of values and morals, perhaps because the child must be preoccupied with the demands of social discipline. The child must know of every act whether it will fall in the category of the approved or the disapproved, the allowable or the forbidden, the right or the wrong. Shades of gray in the moral world are so unsettling that even in adolescence, the young will pressure adults into maintaining the pretense that the over-simple, dual distinctions apply to all acts. "Never mind the flim flam—do you approve or don't you? Can I or can't I? Do I pass or don't I?"

This view of the world feels perfectly coherent from the inside. I go as a child to my parent or my teacher with my little offering for judgment—my wish to go play with my neighbor, my spelling word, my arithmetic example—is it right, or wrong? The large beings before me survey a platonic idea of rightness known to them, or perhaps engraved on some tablet in the sky, and inform me, "Yes, it is right," or "No, it does not conform." In such a world, divided down the middle, it behooves me to stay as much as possible on the side of right. In school, instances on the side of right and wrong, correct and incorrect, are all recorded, added algebraically, and the result used to indicate the degree to which I have been right or wrong, good or bad. In this total I am periodically compared with my classmates above and below me, on a single scale of goodness—of success and of failure.

This picture of the world receives, of course, some severe jolts. Mother may say "Yes" and Father "No." In high school, I find that English teachers disagree about the value and even the meaning of certain poems. At first, disillusion makes me suspect the competence of my particular teachers, but I find that others are no better. Perhaps the

---

* Perry, William G. (1970) Forms of Intellectual and Ethical Development in the College Years: A Scheme. New York: Holt, pp. 28-40.

tablets of truth about poems are at too high an altitude for anyone to discern, or the sky over English teachers is particularly cloudy. If so, then, at least I am free. I suddenly see that the world is not as I first thought, divided between right and wrong. No, it is divided between those things about which opinions can be determined to be right or wrong and those things about which "anyone has a right to his own opinion" (2). That I continue to be graded on my opinions about poems, I ascribe to the unfairness of English teachers, an unfairness understandable in the light of their desire to hold up their heads in company with teachers of mathematics and physics, who, after all, can know what they're doing.

By this maneuver I have saved the clear dual nature of my world, the only world in which I can demand of authority that it state its rules precisely and abide by them. If English teachers cannot make these rules clear, I can then forget the material, "find out what they like," and give it to them. This is less than high-minded, of course, but they started it.

I still have no real doubts that "right answers" themselves are a matter of morals, not of utility. I see evidence of this everywhere. It is not enough that an answer works; to be really right it must be properly arrived at, that is, by hard work. The same answers come upon by unconventional means, "guesswork" or other cleverness, are counterfeit. Sammy, who is known to work very hard for a few answers he gets, is never scolded. Alex, who is discovered to have been exploiting a knack for getting all the answers without cracking a book, is taken aside and talked to for not making the most of his opportunities. The Right still reigns.

The crisis may be harder to postpone in college. In college I am older and stronger and at the same time I am severely shaken by the absence of solidarity among my peers. It may be that other students are in the same predicament, but for the first time I see that they differ from me radically in regard to the things they assign to right or wrong, and to the determinable or the indeterminable. In school the reiteration of the "right" and of authority's limits was the affirmation of friendship's bonds. In the college dormitory it appears that I must cease to reject the "wrong" if I am to have any friends at all.

Worse than this I can no longer maintain the illusion that virtue alone determines rewards in the intellectual world. It is all too clear now that Sammy's efforts may fail to gain him honors and that Alex, who only reads Sammy's notes after sleeping through class, may receive an A. I get discouraged by my hours of labor over themes, which bring me only C; I procrastinate; I guiltily dash off a midnight scrawl—and received B+. The foundations are crumbling.

A heavy burden now falls on my teachers to justify themselves, and the breakthrough may come in my battle with them. At first, the old story repeats itself. I came to college with a new faith in my teachers—perhaps now my Humanities instructor will not be so ignorant of the graven tablets of the truth. Or if the tablets are not visible, he will at least recognize "everyone's right to his own opinion." My hopes are raised when he adjures the class not to write a mere summary of the author's views but to state our own ideas, our own opinions. I do so, my opinion begin largely of the form that I like the book or I don't. I receive a D. Alas, my teacher is not to be trusted after all, and is revealed as a mere section man, wrapped up in his own efforts toward a Ph.D., and hypocritically subservient to those prejudices necessary to minions. The next section meetings will be devoted to elaborate efforts to bait him into the revelation of these prejudices.

As part of this maneuver, I take my paper to him for his comments, to find out "what he wants." He informs me that I state too many generalities with too little data. Suppressing my retort, "But sir, you said . . . ," I leave to fill my next paper with data with a vengeance. The D on this paper is attributed to my lack of ideas. "Look," says the section man, and for a moment we face each other, even across my resentment, "You must learn to show how the facts relate to each other to generate your ideas, to support them. The ideas and the facts must go together, and you must not let them fly in the face of the implication of some other fact to which you do not refer. And furthermore," he says as he sees me to the door, "the privilege of having your ideas respected depends on your presenting them for what they are, not the truth, but interpretations which you prefer among other interpretations. You may not have to spell other interpretations out but you must let your reader know that you are aware that they exist as relevant qualifications of what you have to say."

Something has happened, but it is a matter of pride not to admit it. I trudge off saying, "All right, if *that's* what they want, I'll give it to them." And I sit down with my next paper to "relate" facts and ideas. Perhaps it is at such a moment as this that I become thoroughly involved and after two hours' work suddenly look up to say, "Holy cats, they *do* relate." I can then put forward my interpretation with pride in its integrity. It is this confidence that allows me to afford the realization that the same data might appear in a different light to others and that we must still all stand judgment. Here I can experience my proper conviction that my ideas are (in a new sense) "right," and still speak with humility.

I am fortunately too involved in the "point" I am trying to make in my paper to notice the full implications of the new world I have recognized. But the implications will present themselves to me one by one, forcing their reiterated choice between courage and despair—unless I find some way to shut my eyes. It will be easy enough to see at the outset that interpretations of a book may lie on a range, with those demonstrating the greatest integrity near the center, and others grading off toward either side toward the relatively untenable. Next it would be clear why very different interpretations, from either side, might be assigned the same value.

Soon I may begin to miss those tablets in the sky. If this defines the truth for term papers, how about people? Principalities? Powers? How about the Deity Himself? And if this can be true of my image of the Deity, who then will cleanse my soul? And my enemies? Are they not *wholly* in the wrong?

I apprehend all too poignantly now that in the most fateful decisions of my life I will be the only person with a first-hand view of the really relevant data, and only part of it at that. Who will save me then from that "wrong decision" I have been told not to make lest I "regret-it-all-my-life"? Will no one tell me if I am right? Can I never be sure? Am I alone?

It is not for nothing that the undergraduate turns metaphysician.

Not all students are "sophomores," in this sense, in their sophomore year. Some come to college as "juniors" or even "seniors." Some go all the way through college and somehow manage to remain school-boys to the end. In the sense in which we are speaking, indeed, many people achieve the consequences of a college education without ever going to college at all. The function of a college, however, is to present to the students' attention in concentrated form all the questions that the sophomore in man has raised for himself through the ages and which he has then spent the rest of his history trying to resolve, rephrase, or learn to live with.

We need not stop to analyze the motives that bring man to metaphysics. It seems evident enough that the higher animals gained their relative freedom and mastery through developing the ability to form concepts—that is, to think. Man has gone on to his own greater freedom—and bewilderment—by learning to conceptualize about concepts, to think about his thoughts. Man is distinguished from the ape not by his reason, at which the ape is often no slouch, but by his meta-reason, which is a blessing with which the ape is presumably uncursed. The characteristic of the liberal arts education of today, as we have pointed out, is its demand for a sophistication about one's own line of reasoning as contrasted with other possible lines of reasoning. In short, it demands meta-thinking.

William James would hold that there is nothing useful or good about a meta-thought unless it has useful consequences for a thought that has useful consequences for an action. We agree, and so do most of our students. It is when an intellectual community idealizes its own techniques for their own sake that it forgets action. Meta-meta-thoughts become "higher" than meta-thoughts, and meta-meta-meta-thoughts higher yet, *per astra ad absurdum*.

It is not this tendency that causes the sophomore his pain, even though, when the first floodgates let go he gets lost in just this way.

The issue at hand, we think our records show, is responsibility. If all I have been taught up to now is open to question, especially to *my* question, then my sense of who is responsible shifts radically from outside to me. But I see too that my questions and my answers are likewise open to question. Yet if I am not to spend my life in questions about questions and am to act, choose, decide and live, on what basis am I to do it? I even see now that I have but one life to live.

This then is the issue of individual personal commitment in a relative world, the next step beyond the questions of the "sophomore." Its central burden, and joy, is responsibility. If one quails before it, there are many well trodden paths to postponement, escape, or even retreat. We shall mention these as they appear in our records later. They can be seen most clearly against the experience of those who take up the challenge.

The commitment we are talking about is of a special form. We have called it personal commitment in a relative world. By this we mean to distinguish it from commitments which have been taken for granted to the extent that they have never been questioned, never compared to alternatives which could be "thinkable" to the self. All of us operate on many habitual never-questioned commitments. For some they constitute the entirety of life—that is, the unexamined life. Socrates said such a life was not worth living. Surely this statement is extreme and reveals the snobbery of consciousness. People with unexamined lives have been known to fight for their lives very well indeed. Of course, they could not tell you why and remain people with unexamined lives.

Unexamined commitments can exist in all areas of life. One of our students, diligently "committed" to his goal of medicine ever since he could remember, never asked himself whether he really wanted to be a doctor until he was admitted to medical school. At this point the question hit him like an earthquake. When he had weathered the crisis and decided to go on, his commitment was of the form to which we particularly refer.

In religious life the distinction has long been familiar as the difference between simple belief and faith. Belief may come from one's culture, one's parents, one's habit; faith is an affirmation by the person. Faith can exist only after the realization of the possibility of doubt. We shall have more to say about the relation of religion to the intellectual and emotional growth of our students. We are concerned now with their experience of commitment as we have defined it.

Our students experience all such commitments as affirmations of themselves. Many of our students use the terms of existential philosophy in describing them, though most do so apologetically, knowing the ease with which the jargon can take over. The feeling they describe is one of some decision, some choice among actions, values or meaning which comes from themselves and defines them as individuals. Not that they feel self-created, as if they could choose all their values without reference to their past. On the contrary their commitments seem always to be made in acceptance of their past. Even when a student is breaking with the tradition of his upbringing, he seems first to have to accept the fact of it, that this happened to him, that he lived in it, and that now he must take a stand over against it, knowing that a part of himself must pay the cost.

More usually commitments follow lines similar to those laid down in the unexamined past, but the act of affirmation brings a new and different feel to it. The student who finally "decided" to become a doctor was not unmindful that his long efforts carried weight and momentum into his choice. Yet he did not feel trapped or resigned or passive. He knew he might be fooling himself, but had to take this chance also. He had to decide for himself even about the degree to which he could feel that he had really "decided."

What is required is a capacity for detachment. One must be able to stand back from oneself, have a look, and *then* go back in with a new sense of responsibility.

The act of standing back is forced in a liberal arts college by the impact of pluralism of values and points of view. The stock may be intentional on the part of individual professors, as it is most frequently, though not always, in courses in general Education, or it may be simply the by-product of the clash of different professors, each one of whom is sure he teaches "the" truth. Only in the smallest and most carefully guarded faculties can this diversity be avoided.

We gather from what our students have told us that the educational impact of diversity can be at its best when it is deliberate. When a teacher asks his students to read conflicting authorities and then asks them to assess the nature and meaning of the conflict, he is in a strong position to assist them to go beyond simple diversity into the disciplines of relativity of thought through which specific instances of diversity can be productively exploited. He can teach the relation, the relativism, of one system of thought to another. In short, he can teach disciplined independence of mind.

This is the commonplace of good teaching of the liberal arts in colleges and in good schools of today. And the idea is older than Socrates. In more recent times Henry Adams said that if he were ever to do college lecturing again it would be in the company of an assistant professor whose sole duty would be to present to the students an opposite point of view.

We think we are describing, however, a new thing under the sun. Deliberate teaching of this sort seems to be no longer the exception but the rule—so much the rule that it becomes the very heart of liberal education as revealed in these records. We wonder if this event is not the product of a great educational revolution of the past fifty years.

Some evidence is at hand. On this evidence, the faculty at Harvard appears to have revised its conception of the educated man in the past fifty years. The Harvard faculty may not, of course, be representative of institutions of higher learning in this country. But even in the face of notable instances of leadership by other colleges, we doubt that Harvard has ever been more than ten or fifteen years behind the times in its definition of knowledge in its students.

The faculty's emphasis on independence of thought in examinations coincides, then, with the students' concerns in these records. There is hardly any doubt that the faculty's deliberate effort is a good thing. One would have to be quite anti-rational to maintain that education consciously and thoughtfully considered is less to be desired than that which happens accidentally. But it is not without its pitfalls. Education for independence of mind is a tricky business, as these records show. Unlike the haphazard clash of dogmatic professors, it can double back on itself and undo its own good works.

The problem is not simply that a teacher's bias can sneak back into his efforts to be impartial and subvert his offer of freedom. It often does, but the students soon discover how to deal with it, even when it appears in forms subtle and unlovely. The problem, as these records occasionally document, is that, where independence of mind is demanded by authority, its forms can be mastered and "handed in" while the spirit remains obediently conformist. As a student said of his performance on an examination, "Well, I decided to be in favor of that book they asked about, but I did not forget to be balanced."

The "pros" and "cons," the glib presentation of "several points of view," the summary which judiciously selects one position to be in favor of, "all things considered"—these become the stock armamentarium of the gamesman who has "caught on" to "what they want" and is giving it to them in exchange for grades and a diploma.

From what our students have revealed to us in their candor we find it impossible to imagine an educational system that would be proof against a wish on their part to defeat its ends. They would always find ways of imitating, of holding before the tired eyes of the professor the image of his fondest hope, all done up in his favorite words and his pet references and treasured qualifications. While their souls . . .

Students' souls, we have learned from both these researches and our daily counseling, are safe even against being saved. We find this encouraging. Certainly we prefer to see gamesmanship played cynically, if it is to be played at all, rather than automatically and unconsciously. The cynic, at least, maintains a claim to independence.

Modern pluralistic education, with all its pros and cons on every subject, is criticized for not teaching commitment, indeed for leading students away from it. What we have been saying from our understanding of our records is that: (1) without a clear view of pluralism, commitment as we define it is impossible; and (2) commitment can be provided for and given recognition, but it can never be brought about or forced.

There are too many ways out. As in *Pilgrim's Progress* there are stopping places, Sloughs of Despond, paths that lead aside or back. The first crucial trial is in the student's initial confrontation with multiplicity—with pluralism, ambiguity and contingency. One or two of our students would have none of it at all. Presumably, for them to have answers black and white, clear, known, available was so important emotionally that any other world was intolerable. These students either left college or retreated into a deep reaction. Another student stated in effect that this "many-points-of-view business" was O.K. for other people but not for him! A few others accepted the fact of multiplicity, in the loose sense that "everybody has a right to his own opinion," and struggled against the college's demand that they think further.

Perhaps the most critical point in most of the records comes at the moment where the student has indeed discovered how to think further, how to think relatively and contingently, and how to think about thinking. For here it is up to him in what crucial spirit he is to employ this discovery.

He appears to have a number of options. He may allow this form of thought to be simply "what they want" and assume no responsibility himself. He may put his mind in the service of this opportunism and become a cynical gamesman. He may isolate his discovery in the world of academics alone and never allow it to raise questions about his own life and purposes. Or he may see clearly enough what is now incumbent upon him and yet not feel up to it. He can feel not quite old enough or strong enough to make his commitments, and simply procrastinate.

Or he may go on. If he does so, his first commitment may well be to his major, his field of concentration. He has had enough of too much "breadth." He sees now the breadth of his ignorance; it is time to take the plunge, to know some one thing well. And so as a junior he works at his peak and looks forward to his thesis.

Not infrequently, he discovers later, the old hope for certainty has just gone underground, and in his senior year he has the whole job to do over again. "Knowing one thing well" turns out to have meant "Knowing something for sure," and when his thesis, that looked so narrow and specialized, opens out on him into the uncertainties of the whole universe again, he is taken aback.

We think this is the most crucial moment in higher education. Here the student *has* his tools; he has learned how to compare "models" of thought, how to relate data and frames of reference and points of observation. But now differences of opinion among experts in his field appear even more irreconcilable than ever. No one can *ever* be sure. "It's all up to the individual in the end."

Well, indeed it is. But the tone of this statement too often implies "So why bother. If it's all a matter of opinion in the end, why not in the beginning? Why bother with all the intellectual effort?" This is a retreat to simple multiplicity, to "everyone has a right to his own opinion." It says that unexamined opinion is as good as examined opinion. It is the moral defeat of the "educated" man.

It was our wonder and delight, therefore, to hear most of our students survive this crisis also. Many would laugh when they came to the realization that a commitment must be made and remade in time and at deeper levels. They would see also that many areas remained where they had not yet begun to take their stands. But they would go on.

Perhaps the reader will find his reading of a few records more meaningful if we spell out briefly here what we think we have learned from the whole series of records about the anatomy of commitments and the way they go together to make a style of life.

Students seem to think of their commitments in two ways which we will call *area* and *style*. Area refers primarily to social content: decisions regarding major field, career, religion, tastes, morality, politics, finances, social concerns, friendships, and marriage.

The *stylistic* aspects of commitments concern balances both external and subjective. The external balances concern decisions of emphasis between studies and extracurricular activities, number and intensity of friendships, amounts of altruistic social service, degree of specialization vs. breadth and so on. Issues among these external stylistic balances are closely interwoven with commitments in given areas, often determining them or affected by them.

A student's subjective style in regard to his commitments also appears to be both an important part of them and influenced by them. This subjective style involves certain inner commitments or affirmations or acceptances. In these records the students speak often of such inner balances as those between action and contemplation ("I've come to learn when to say to myself 'well, now, enough of this mulling and doubting, let's do something'"); permanence and flexibility ("Sometimes you have to go into something with the feeling you'll be in it forever even when you know you probably won't be" or "I used to think I had to finish anything I started or I'd be a quitter; now I see it's a nice point when to stop something that may be unprofitable and put the effort in more hopeful directions"); control vs. openness ("Well, you have to let experience teach you; it's good to have a plan, but if you insist on following it without a change ever—"); intensity vs. perspective ("That's the trick, I guess, you have to have detachment, or you get lost, you can't see yourself and your relation to what you're doing, and yet if you stay detached you never learn from total involvement, you never live; you just have to do it by waves, I guess").

It is the particular equilibrium that a student finds for himself among such subjective polarities that define him in his feeling of who he is as much as the concrete commitments he makes in different areas of his life. There are others too—a sense of continuity of self through mood and time in the face of the need to take one's immediate feelings seriously; and most particularly the realization that however much one wants to feel one has "found oneself" one wants more to keep growing all one's life ("I sometimes wonder how many times I'll be confronted").

With some moving, flowing equilibrium, some kind of style, among all these issues, the senior in our records tends to look more outward. His competence assured, he tends to be less preoccupied with the ingredients of self-hood, to accept himself as he is and grows, and to hanker for action. In a curious way he may startlingly resemble himself as a freshman. Here is the promising freshman-scientist who has established his competence, is accepted at graduate school and knows just where he's headed. Here is the once freshman-who-would-be-doctor and who next year will be called Doctor "on rounds." They look ahead and outward. They may find little to say in interviews except to allow in a quiet aside that they are getting married next week. Ironically enough these seniors have, some of them, forced us to consider what the difference may be between their kind of outwardness and that of the most hard-shelled anti-intellectual. The difference defines a liberal education—not as an ideal, but as an actuality in real people.

The difference is surely not simply the "content" of so many courses in Chemistry or History. Anti-intellectuals have been known to master mountains of data and technology. The anti-intellectual cannot be passed off as one who refused to think. Many think dangerously well. Similarly the liberally educated man cannot be caricatured as one who sees so many sides of a subject that he cannot act. Our records belie such stereotypes.

We have come to believe from all these hours of listening that the anti-intellectual, be he in or out of college, is definable not as "against thinking," but against thinking about one particular thing: thought. Most particularly his *own* thought.

In contrast, the liberally educated man, be he a graduate of college or not, is one who has learned to think about even his own thoughts, to examine the way he orders his data and the assumptions he is making, and to compare these with other thoughts that other men might have. If he has gone the whole way, as most of our students have done, he has realized that he thinks this way not because his teachers ask him to but because this is how the world "really is," this is man's present relation to the universe. From this position he can take responsibility for his own stand and negotiate—with respect—with other men.

Literature Cited

1. From the *Judge's Manual.*

2. This concept of pluralism will be referred to in these researches as "Multiplicity."

Rare Bird*

Andrea Barrett

Imagine an April evening in 1762. A handsome house set in the gently rolling Kent landscape a few miles outside the city of London; the sun just set over blue squill and beech trees newly leafed. Inside the house are a group of men and a single woman: Christopher Billopp, his sister Sarah Anne, and Christopher's guests from London. Educated and well-bred, they're used to a certain level of conversation. Just now they're discussing Linnaeus's contention that swallows retire under water for the winter - that old belief, stemming from Aristotle, which Linnaeus still upholds.

"He's hardly alone," Mr. Miller says. Behind him, a large mirror reflects a pair of portraits: Christopher and Sarah Anne, painted several years earlier as a gift for their father. "Even Klein, Linnaeus's rival, agrees. He wrote that a friend's mother saw fisherman bring out a bundle of swallows from a lake near Pilaw. When the swallows were placed near a fire, they revived and flew about."

Mr. Pennant nods. "Remember the reports of Dr. Colas? Fishermen he talked to in northern parts claimed that when they broke through the ice in winter they took up comatose swallows in their nets as well as fish. And surely you remember reading how Taletini of Cremona swore a Jesuit had told him that the swallows in Poland and Moravia hurled themselves into cisterns and wells come autumn."

Mr. Collinson laughs at this, although not unkindly, and he looks across the table at his old friend Mr. Ellis. "Hearsay, hearsay," he says. He has a spot on his waistcoat. Gravy, perhaps. Or cream. "Not one shred of direct evidence. Mothers, fishermen, itinerant Jesuits—this is folklore, my friends. Not science."

At the foot of the table, Sarah Anne nods but says nothing. Pennant, Ellis, Collinson, Miller: all distinguished. But old, so old. She worries that she and Christopher are growing prematurely old as well. Staid and dull and entirely too comfortable with these admirable men, whom they have known since they were children.

Their father, a brewer by trade but a naturalist by avocation, had educated Christopher and Sarah Anne together after their mother's death, as if they were brothers. The three of them rambled the grounds of Burdem Place, learning the names of the plants and birds. Collinson lived in Peckham then, just a few miles away, and he often rode over bearing rare plants and seeds sent by naturalist friends in other countries. Peter Kalm, Linnaeus's famous student, visited the Billopps; Linnaeus himself, before Sarah Anne was born, once stayed for several days.

All these things are part of Sarah Anne's and Christopher's common past. And even after Christopher's return from Cambridge and their father's death, for a while they continued to enjoy an easy exchange of books and conversation. But now all that has changed. Sarah Anne inherited her father's brains but Christopher inherited everything else, including his father's friends. Sarah Anne acts as hostess to these men, at Christopher's bidding. In part she's happy for their company, which represents her only intellectual companionship. In part she despises them for their lumbago and thinning

---

hair, their greediness in the presence of good food, the stories they repeat about the scientific triumphs of their youth, and the fact that they refuse to take her seriously. Not one of them has done anything original in years.

There's another reason, as well, why she holds her tongue on this night. Lately, since Christopher has started courting Miss Juliet Colden, he's become critical of Sarah Anne's manners. She does not dress as elegantly as Juliet, or comport herself with such decorum. She's forward when she ought to be retiring, he has said, and disputatious when she should be agreeable. He's spoken to her several times already: "You should wear your learning modestly," he lectures.

She does wear it modestly, or so she believes. She's careful not to betray in public those subjects she knows more thoroughly than Christopher. Always she reminds herself that her learning is only book-learning; that it hasn't been tempered, as Christopher's has, by long discussions after dinner and passionate arguments in coffeehouses with wiser minds.

And so here she is: learned, but not really; and not pretty, and no longer young: last month she turned twenty-nine. Old, old, old. Like her company. She knows that Christopher has begun to worry that she'll be on his hands for life. And she thinks that perhaps he's mentioned this worry to his friends.

They're fond of him, and of Burdem Place. They appreciate the library, the herbarium, the rare trees and shrubs outside, the collections in the specimen cabinets. They appreciate Sarah Anne as well, she knows. Earlier, they complimented the food, her gown, the flowers on the table and her eyes in the candlelight. But what's the use of that sort of admiration? Collinson, who has known her the longest, was the only one to make a stab at treating her the way they all had when she was a girl: he led her into quoting Pliny and then complimented her on her learning. But she saw the way the other men shifted uneasily as she spoke.

Despite herself, she continues to listen to the men's conversation. Despite her restlessness, her longing to be outside in the cool damp air, or in some other place entirely, she listens because the subject they're discussing fascinates her.

"I had a letter last year from Solander," Ellis says. "Regarding the November meeting of the Royal Society. There, a Reverend Forster said he'd observed large flocks of swallows flying quite high in the autumn, then coming down to sit on reeds and willows before plunging into the water of one of his ponds."

"More hearsay," Collinson says.

But Pennant says it might be so; either that or they slept for the winter in their summer nesting holes. "Locke says that there are no chasms or gaps in the great chain of being," he reminds them. "Rather there is a continuous series in which each step differs very little from the next. There are fishes that have wings, and birds that inhabit the water, whose blood is as cold as that of fishes. Why should not the swallow be one of those animals so near of kin to both birds and fish that it occupies a place between both? As there are mermaids or seamen, perhaps."

No one objects to the introduction of aquatic anthropoids into the conversation. Reports of them surface every few years—Cingalese fishermen swear they've caught them in their nets, a ship's captain spots two off the coast of Massachusetts. In Paris, only four years ago, a living female of the species was exhibited.

Collinson says, "Our friend Mr. Achard writes me that he has seen them hibernating in the cliffs along the Rhine. But I have my doubts about the whole story."

"Yes?" Pennant says. "So what do you believe?"

"I think swallows migrate," Collinson says.

While the servants change plates, replace glasses, and open fresh bottles of wine, Collinson relates a story from Mr. Adanson's recent *History of Senegal*. Off the coast of that land in autumn, he says, Adanson reported seeing swallows settling on the decks and rigging of passing ships like bees. Others have reported spring and autumn sightings of swallows in Andalusia and over the Strait of Gibraltar. "Clearly," Collinson says, "they must be birds of passage."

Which is what Sarah Anne believes. She opens her mouth and proposes a simple experiment to the men. "The swallow must breathe during winter," she says, between the soup and the roasted veal. "Respiration and circulation must somehow continue, in some degree. And how is that possible if the birds are under water for so long? Could one not settle this by catching some swallows at the time of their autumn disappearance and confining them under water in a tub for a time? If they are taken out alive, then Linnaeus's theory is proved. But if not . . ."

"A reasonable test," Collinson says. "How would you catch the birds?"

"At night," she tells him impatiently. Oh, he is so old; he has dribbled more gravy on his waistcoat. How is that he can no longer imagine leaving his world of books and talk for the world outside? Anyone might gather a handful of birds. "With nets, while they roost in the reeds."

Collinson says, "If they survived, we might dissect one and look for whatever internal structure made possible their underwater sojourn."

He seems to be waiting for Sarah Anne's response, but Christopher is glaring at her. She knows what he's thinking: in his new, middle-aged stodginess, assumed unnecessarily early and worn like a borrowed coat, he judges her harshly. She's been forward in entering the conversation, unladylike in offering an opinion that contradicts some of her guests, indelicate in suggesting that she might pursue a flock of birds with a net.

What has gotten into him? That pulse she hears inside her ear, the steady swish and hum of her blood, is the sound of time passing. Each minute whirling past her before she can wring any life from it; hours shattered and lost while she defers to her brother's sense of propriety.

Upstairs, finally. Dismissed while the men, in the library below, drink Christopher's excellent wine and avail themselves of the chamberpot in the sideboard. Her brother's friends are grateful for her hospitality, appreciative of her well-run household; but most grateful and appreciative when she disappears.

Her room is dark, the night is cool, the breeze flows through her windows. She sits in her high-ceilinged room, at the fragile desk in the three-windowed bay facing west, over the garden. If it were not dark, she could see the acres leading down to the lake and the low stretch of rushes and willows along the banks.

Her desk is very small, meant to hold a few letters and a vase of flowers: useless for any real work. The books she's taken from the library spill from it to the floor. Gorgeous books, expensive books. Her brother's books. But her brother doesn't use them the way she does. She's been rooting around in them and composing a letter to Linnaeus, in Uppsala, about the evening's dinner conversation. Christopher need never know what she writes alone in this room.

Some years ago, after Peter Kalm's visit, Sarah Anne's father and Linnaeus corresponded for a while; after expressing admiration for the great doctor's achievements this visit is what she first mentions. Some flattery, some common ground. She discusses the weather, which has been unusual; she passes on the news of Collinson's latest botanical acquisitions. Only then does she introduce the subject of the swallows. She writes:

> *Toward the end of September, I have observed swallows gathering in the reeds along the Thames. And yet, although these reeds are cut down annually, no one has ever discovered swallows sleeping in their roots, nor has any fisherman ever found, in the winter months, swallows sleeping in the water. If all the great flocks seen in the autumn dove beneath the water, how could they not be seen? How could none be found in winter? But perhaps the situation differs in Sweden.*
>
> *You are so well-known and so revered. Could you not offer the fishermen of your country a reward, if they were to bring to you or your students any swallows they found beneath the ice? Could you not ask them to watch the lakes and streams in spring, and report to you any sightings of swallows emerging from the water? In this fashion you might elucidate the problem.*

She pauses and stares at the candle, considering what she observed last fall. After the first killing frosts, the swallows disappeared along with the warblers and flycatchers and other insectivorous birds deprived of food and shelter. Surely it makes sense that they should have gone elsewhere, following their food supply?

She signs the letter "S. A. Billopp," meaning by this not to deceive the famous scholar but simply to keep him from dismissing her offhand. Then she reads it over, seals it, and snuffs her candle. It is not yet ten but soon the men, who've been drinking for hours, will be expecting her to rejoin them for supper. She will not go down, she will send a message that she is indisposed.

She rests her elbows on the windowsill and leans out into the night, dreaming of Andalusia and Senegal and imagining that twice a year she might travel like the swallows. Malaga, Tangier, Marrakech, Dakar. Birds of passage fly from England to the south of France and from there down the Iberian peninsula, where the updrafts from the Rock of Gibraltar ease them over the Strait to Morocco. Then they make the long flight down the coast of Africa.

A bat flies by, on its way to the river. She has seen bats drinking on the wing, as swallows do, sipping from the water's surface. Swallows eat in flight as well, snapping insects from the air. Rain is sure to follow when they fly low; a belief that dates from Virgil, but which she knows to be true. When the air is damp and heavy the insects hover low, and she has seen how the swallows merely follow them.

In the dark she sheds her gown, her corset, her slippers and stockings and complicated underclothes, until she is finally naked. She lies on the floor beside her desk, below the open window. Into her notebook she has copied these lines, written by Olaus Magnus, archbishop of Uppsala, in 1555:

> *From the northern waters, swallows are often dragged up by fishermen in the form of clustered masses, mouth to mouth, wing to wing, and foot to foot, these having at the beginning of autumn collected amongst the reeds previous to submersion. When young and inexperienced fisherman find such clusters of swallows, they will, by thawing the birds at the fire, bring them indeed to the use of their wings, which will continue but a very short time, as it is a premature and forced revival; but the old, being wiser, throw them away.*

A lovely story, but surely wrong. The cool damp air washes over her like water. She folds her arms around her torso and imagines lying at the bottom of the lake, wings wrapped around her body like a kind of chrysalis. It is cold, it is dark, she is barely breathing. How would she breathe? Around her are thousands of bodies. The days lengthen, some signal arrives, she shoots with the rest of her flock to the surface, lifts her head and breathes. Her wings unfold and she soars through the air, miraculously dry and alive.

It is possible?

Eight months later, Sarah Anne and Christopher stand on London Bridge with Miss Juliet Colden and her brother John, all of them wrapped in enormous cloaks and shivering despite these. They've come to gaze at the river, which in this January of remarkable cold is covered with great floes of ice. An odd way, Sarah Anne thinks, to make the announcement of Christopher and Juliet's engagement. She wishes she liked Juliet better. Already they've been thrown a great deal into each other's company; soon they'll be sharing a house.

But not sharing, not really. After the wedding, Juliet will have the household keys; Juliet will be in charge of the servants. Juliet will order the meals, the flowers, the servants' livery, the evening entertainments. And Sarah Anne will be the extra woman.

The pieces of ice make a grinding noise as they crash against each other and the bridge. Although the tall brick houses that crowded the bridge in Sarah Anne's childhood were pulled down several years ago and no longer hang precariously over the water, the view remains the same: downriver the Tower and a forest of masts; upriver the Abbey and Somerset House. The floating ice greatly menaces the thousands of ships waiting to be unloaded in the Pool. It is of this that John and Christopher speak. Manly talk: will ships be lost, fortunes destroyed? Meanwhile Juliet chatters and Sarah Anne is silent, scanning the sky for birds.

Wrynecks, white-throats, nightingales, cuckoos, willowwrens, goatsuckers—none of these are visible, they've disappeared for the winter. The swallows are gone as well. An acquaintance of Christopher's mentioned over a recent dinner that on a remarkably warm December day, he'd seen a small group of swallows huddled under the moldings of a window at Merton College. What were they doing there? She's seen them, as late as October, gathered in great crowds in the osier-beds along the river—very late for young birds attempting to fly past the equator. In early May she's seen them clustered on the largest willow at Burdem Place, which hangs over the lake. And in summer swallows swarm the banks of the Thames below this very bridge. It's clear that they're attached to water, but attachment doesn't necessarily imply habitation. Is it possible that they are still around, either below the water or buried somehow in the banks?

If she were alone, and not dressed in these burdensome clothes, and if there were some way she could slip down one of the sets of stairs to the river bank without arousing everyone's attention, she knows what she would do. She'd mark out a section of bank where the nesting holes are thickest and survey each hole, poking down the burrows until she found the old nests. In the burrows along the river bank at home she's seen these: a base of straw, then finer grass lined with a little down. Small white eggs in early summer. Now, were she able to look, she believes she'd find only twists of tired grass.

The wind blows her hood over her face. As soon as she gets home, she thinks, she'll write another letter to Linnaeus and propose that he investigate burrows in Sweden. Four times she's written him, this past summer and fall; not once has he answered.

Christopher and John's discussion has shifted to politics, and she would like to join them. But she must talk to Juliet, whose delicate nose has reddened. Juliet's hands are buried in a huge fur muff; her face is buried in her hood. Well-mannered, she refuses to complain of the cold.

“You'll be part of the wedding, of course," Juliet says, and then she describes the music she hopes to have played, the feast that will follow the ceremony. "A big table," she says. "On the lawn outside the library, when the roses are in bloom—what is that giant vine winding up the porch there?"

"Honeysuckle," Sarah Anne says gloomily.

She can picture the wedding only too clearly. The other attendants will be Juliet's sisters, all three as dainty and pretty as Juliet. Their gowns will be pink or yellow or pink and yellow, with bows down the bodice and too many flounces. The couple will go to Venice and Paris and Rome and when they return they'll move into Sarah Anne's large sunny bedroom and she'll move to a smaller room in the north wing. The first time Juliet saw Sarah Anne's room, her eyes lit with greed and pleasure. A few days later Christopher said to Sarah Anne, "About your room . . ." She offered it before he had to ask.

"Christopher and I thought you'd like the dressing table your mother used," Juliet says. "For that lovely bay in your new room."

But just then, just when Sarah Anne thinks she can't bear another minute, along comes another of her dead father's elderly friends, accompanied by a woman. Introductions are made all around. Mr. Hill, Mrs. Pearce. Sarah Anne has always enjoyed Mr. Hill, who is livelier than his contemporaries, but he is taken away. The group splits naturally into two as they begin their walk back to the Strand. Mr. Hill joins Christopher and John, and Mrs. Pearce joins Sarah Anne and Juliet. But Mrs. Pearce, instead of responding to Juliet's remarks about the weather, turns to Sarah Anne and says, "You were studying the riverbank so intently when Mr. Hill pointed you out to me. What were you looking for?"

Her face is lean and intelligent; her eyes are full of curiosity. "Birds," Sarah Anne says impulsively. "I was looking for swallows' nests. Some people contend that swallows spend the winter hibernating either under water or in their summer burrows."

She explains the signs that mislead observers, the mistaken stories that multiply. At Burdem Place, she says, she heard a friend of her brother's claim that, as a boy, he found two or three swallows in the rubble of a church-tower being torn down. The birds were torpid, appearing dead, but revived when placed near a fire. Unfortunately they were then accidentally roasted.

"Roasted?" Mrs. Pearce says with a smile.

"Crisp as chickens," Sarah Anne says. "So of course they were lost as evidence. But I suppose it's more likely that they overwinter in holes or burrows, than that they should hibernate under water."

"Some people read omens in the movements of swallows," Mrs. Pearce says. "Even Shakespeare—remember this? 'Swallows have built in Cleopatra's sails their nests. The auguries say they know not, they cannot tell, look grimly, and dare not speak their knowledge.' Poetic. But surely we're not meant to believe it literally."

Sarah Anne stares. There's nothing visibly outrageous about Mrs. Pearce. Her clothing is simple and unfashionable but modest; her hair is dressed rather low but not impossibly so. "I believe that one should experiment," Sarah Anne says. "That we should base our statements on evidence."

"I always prefer to test hypotheses for myself," Mrs. Pearce says quietly.

Juliet is pouting, but Sarah Anne ignores her. She quotes Montaigne and Mrs. Pearce responds with a passage from Fontenelle's *Entretiens sur la pluralite des mondes*. "Do you know Mrs. Behn's translation?" Sarah Anne asks. At that moment she believes in a plurality of worlds as she never has before.

"Of course," says Mrs. Pearce. "Lovely, but I prefer the original."

Sarah Anne mentions the shells that she and Christopher have inherited from Sir Hans Sloane's collection, and Mrs. Pearce talks about her collection of mosses and fungi. And when Sarah Anne returns to the swallows and says that Linnaeus's belief in their watery winters derives from Aristotle, Mrs. Pearce says, "When I was younger, I translated several books of the 'Historia Animalium.'"

Sarah Anne nearly weeps with excitement and pleasure. How learned this woman is. "How were you educated?" she asks.

"My father," Mrs. Pearce says. "A most cultured and intelligent man, who believed girls should learn as well as their brothers. And you?"

"Partly my father, partly my brother, before . . . Partly by stealth."

"Well, *stealth*," Mrs. Pearce says with a little smile. "Of course."

In their excitement they've been walking so fast that they've left Juliet behind. They hear the men calling them and stop. Quickly, knowing she has little time, Sarah Anne asks the remaining important questions. "And your husband?" she says. "He shares your interests?"

"He's dead," Mrs. Pearce says calmly. "I'm a widow."

She lives in London, Sarah Anne learns, alone but for three servants. Both her daughters are married and gone. "I would be so pleased if you would visit us," Sarah Anne says. "We have a place just a few miles from town, but far enough away to have all the pleasures of the country. In the gardens there are some interesting plants from North America, and we've quite a large library . . ."

Mrs. Pearce lays her gloved hand on Sarah Anne's arm. "I'd be delighted," she says. "And you must visit me in town. It's so rare to find a friend."

The others join them, looking cold and displeased. "Miss Golden," Mrs. Pearce says.

"Mrs. Pearce. I do hope you two have had a nice talk."

"Lovely," Mrs. Pearce says.

She looks over Juliet's head at Sarah Anne. "I'll see you soon." Then she hooks her hand into Mr. Hill's arm and walks away.

"Odd woman," John says. "Bit of a bluestocking, isn't she?"

"She dresses terribly," Juliet says with considerable satisfaction. From the sharp look she gives Sarah Anne, Sarah Anne knows she'll pay for that brief bit of reviving conversation. But her mind is humming with the pleasure of her new friend, with plans for all they might do together, with the letter she'll write to Linnaeus the very instant she reaches home. She imagines reading that letter out loud to Mrs. Pearce, showing Mrs. Pearce the response she will surely receive.

* * * *

"We should write him about that old potion," Mrs. Pearce says; and Sarah Anne says, "What?"

"For melancholy. Don't you know it?"

"I don't think so."

"It's a potion made partly from the blood of swallows. Birds of summer, symbols of ease—the potion is supposed to ease sadness and give wings to the feet."

"More likely than what he's proposing," Sarah Anne says, and Mrs. Pearce agrees.

It's September now—not the September following their meeting but the one after that: 1764. The two women are in an unused stable at Burdem Place, patiently waiting, surrounded by their equipment. It is just barely dawn. Down in the reeds, where the birds are sleeping, they've seen Robert the gardener's boy with a net and instructions. What they're talking about while they wait is the letter Sarah Anne received last week from Carl Linnaeus, in which he graciously but firmly (and in Latin; but Sarah Anne can read it), dismissed her theories and stated his absolute conviction that swallows hibernate under the water. The letter upset Sarah Anne, but she would not have done anything more than fume had Mrs. Pearce not been visiting. It was Mrs. Pearce—Catherine—who said, "Well. We'll just have to do the experiments ourselves."

On the wooden floor they've set the bottom half of a cask, which Robert has filled with water. Below the water lies a few inches of river sand; on the surface a board floats an inch from the rim. A large piece of sturdy netting awaits the use of which they'll put it. Inside the stable it's still quite dark; through the open door the trees are barely visible through the mist. Above them the house sleeps. Just after four o'clock, Sarah Anne rose in her new room and tapped once on the door of the room down the hall, where Catherine stays when she visits. Catherine opened the door instantly, already dressed.

Recently it has been easier for them to talk about the swallows than about the other goings-on at Burdem Place. Juliet's pregnancy has made her ill-humoured, and Christopher has changed as well. Sarah Anne knows she should have expected this, but still it has come as a shock. These days the guests tend to be Juliet's frivolous friends and not the older naturalists. Young, not old; some of them younger than Sarah Anne herself. For weeks at a time they stroll the grounds in fancy clothes and play games while Sarah Anne hovers off to the side, miserable in their company.

Who is she, then? She doesn't want to act, as Christopher does, the part of her parents' generation; but now she's found that she doesn't like her own peers either. She fits nowhere. Nowhere, except with Catherine. She and Catherine, tucked into a wing away from the fashionable guests, have formed their own society of two. But she suspects that, after the birth of Juliet's child, even this will be taken from her.

Christopher hopes for many children, an army of children. This child, and the ones that follow, will need a nurse and a governess, Juliet says. And a nursery, and a schoolroom. Sarah Anne has seen Christopher prowling the halls near her bedroom, assessing the space and almost visibly planning renovations. He's welcomed Catherine's frequent long visits—but only, Sarah Anne knows, because they keep her occupied and him from feeling guilty about her increasing isolation. The minute he feels pinched for space, he'll suggest to Sarah Anne that Catherine curtail her visits. And then it's possible he'll ask Sarah Anne to be his children's governess.

But Sarah Anne and Catherine don't talk about this. Instead they look once more at Linnaeus's letter, which arrived addressed to "Mr. S. A. Billopp" but which, fortunately, Christopher didn't see. They arrange their instruments on the bench beside them and shiver with cold and excitement. They wait. Where is Robert?

It was Catherine who first approached this weedy twelve-year-old, after Sarah Anne told her she'd once overhead him talking about netting birds for food in Ireland. Catherine told him that they required two or three swallows and would pay him handsomely for them; Robert seemed to believe they had plans to eat them. Still, at 4:30 he met them here, silent and secret. Now he reappears in the doorway, barefoot and wet to the waist. His net is draped over one shoulder and in his hands he holds a sack, which pulses and moves of its own accord.

"Robert!" Catherine says. "You had good luck?"

Robert nods. Both his hands are tightly wrapped around the sack's neck and when Catherine reaches out for it he says, "You hold this tight, now. They'll be wanting to fly."

"You did a good job," Catherine says. "Let me get your money. Sarah Anne, why don't you take the sack?"

Sarah Anne slips both her hands below Robert's hands and twists the folds of cloth together. "I have it," she says. Robert releases the sack. Immediately she's aware that the sack is alive. Something inside is moving, leaping, dancing. Struggling. The feeling is terrifying.

"Thank you, Robert," Catherine says. Gently she guides him out the door. "You've been very helpful. If you remember to keep our secret, we'll ask you for help again."

By the time she turns back to Sarah Anne and takes the sack from her, Sarah Anne is almost hysterical.

"Nothing can satisfy but what confounds," Catherine says. "Nothing but what astonishes is true." Once more Sarah Anne is reminded of her friend's remarkable memory. When Catherine is excited, bits of all she has ever read fly off her like water from a churning lump of butter.

"All right now," Catherine says. "Hold the netting in both hands and pull it over the tub - that's good. Now fasten down the sides, all except for this little section here. I'm going to hold the mouth of the sack to the open part of the netting, and when I say the word I'll open the sack and you drop the last lip of the netting into place. Are you ready?"

"Ready," Sarah Anne says. Her heart beats if she has a bird inside her chest.

"*Now*," Catherine says.

Everything happens so fast—a flurry of hands and cloth and netting and wings, loops of string and snagged skirts. Two swallows get away, passing so close to Sarah Anne's face that she feels the tips of their feathers and screams. But a minute later she sees that they've been at least partly successful. In the tub, huddled on the board and pushing frantically at the netting, are two birds. Steely blue, buff-bellied, gasping.

"They're so unhappy," Sarah Anne says.

"We must leave them," Catherine says. 'If the famous Doctor Linnaeus is right, in our absence they'll let themselves down into the water and sleep, either on the surface of the river sand or perhaps just slightly beneath it."

"And if he's wrong?"

"Then we'll tell him so."

The day passes with excruciating slowness, chopped into bits by Juliet's rigid timetable: family breakfast, dinner, tea, and supper, long and complicated meals. After breakfast Juliet requires the company of Sarah Anne and Catherine in her dressing room, although Sarah Anne knows that Juliet is fond of neither of them. After tea, Christopher expects the women to join him in the library, where they talk and read the newspapers. Sarah Anne and Catherine have not a minute to themselves, and by supper they're wild-eyed with exhaustion and anticipation.

The next morning, when they slip out again before breakfast, the board over the tub is bare. Sarah Anne unfastens the netting, removes the dripping board, and peers down into the water. The swallows lie on the sand. But not wrapped serene in a cocoon of wings; rather twisted and sprawled. She knows before she reaches for them that they're dead. Catherine knows too; she stands ready with a penknife. They've agreed that, should the swallows die, they'll dissect one and examine its structures of circulation and respiration. They'll look for any organ that might make hibernation under water possible; any organ that might prove them wrong.

They work quickly. There isn't much blood. Catherine, peering into the open chest cavity, says, "It is very difficult to work without proper tools. Still. There is nothing out of the ordinary here. And there is no doubt that Linnaeus is wrong."

A four-chambered heart inside its pericardium; small, rosy, lobeless lungs. From the lungs, the mysterious air sacs extend into the abdomen, up into the neck, into the bones. There is no sign of a gill-like organ that might allow the bird to breathe under water. Sarah Anne is quite faint, and yet also fiercely thrilled. They've done an experiment; they've disproved an hypothesis. She says, "We will write to Linnaeus today."

"I think not," Catherine says. "I think it's time we make other plans."

What plans were those? Of course Christopher noticed that Mrs. Pearce returned to London in early October; he noticed, too, when Sarah Anne left Burdem Place a few weeks later for what she described as an 'extended visit' with her friend. All through November Christopher didn't hear from his sister, but he had worries of his own and thought nothing of her absence. In December, when he was in London on business, he stopped by Mrs. Pearce's house to find that her servants had been dismissed and her house was empty. Only then did he realize that his sister and her friend were simply gone.

Everyone had theories about their disappearance: Collinson, Ellis, all the men. Foul play was suspected by some, although there was no evidence. But this is what Christopher thought, during the bleak nights of 1765 while Juliet was writhing with childbed fever, and during the even bleaker nights after her death, while his tiny son was wasting away. He imagined Sarah Anne and Mrs. Pearce—and who was Mrs. Pearce anyway? Where had she come from? Who were her people? –up before dawn in that London house, moving swiftly through the shadows as they gather bonnets, bags, gloves. Only one bag apiece, as they mean to travel light: and then they glide down the early morning streets toward the Thames. Toward the Tower wharf, perhaps; but it could be any wharf, any set of stairs, the river hums with activity. Ships are packed along the waterfront, their sails furled and their banners drooping; here a wherry, there a cutter, darts between them and the stairs. Some of the ships are headed for India and some for Madagascar. Some are going to the West Indies and others to Africa. Still others are headed for ports in the North American provinces: Quebec or Boston, New York or Baltimore.

Christopher believes his sister and her companion have boarded one of the ships headed for America. Once he overhead the two of them waxing rhapsodic over Mark Catesby's *Natural History*, talking in hushed tones about this land where squirrels flew and frogs whistled and birds the size of fingernails swarmed through forests so thick the sunlight failed to reach the ground. Catesby, Sarah Anne said, believed birds migrated sensibly: they flew to places where there was food.

Pacing his lonely house, miserable and broken, Christopher imagines the ship slowly moving down the Thames toward Dover and the Channel. There's a headwind and the tides are against them; the journey to Dover takes three days. But then the wind shifts and the luck arrives. They fly past Portsmouth and Plymouth and Land's End, into the open ocean. The canvas billows out from the spars; the women lean against the railings, laughing. That was the vision he had in mind when, a few years later, he sold both Burdem Place and the brewery and sailed for Delaware.

He never found Sarah Anne. But the crossing and the new world improved his spirits; he married a sturdy young Quaker woman and started a second family. Among the things he brought to his new life were two portraits—small, sepia-toned ovals, obviously copies of larger paintings—which surfaced much later near Baltimore. And if the faded notes found tucked in the back of Christopher's portrait are true, he made some modest contributions to the natural history of the mid-Atlantic states.

Sarah Anne's portrait bears only the date of her birth. Her letters were discovered in the mid-1850s, in the attic of a distant relation of the husband of Linnaeus's youngest daughter, Sophia. The British historian who found them was editing a collection of Linnaeus's correspondence, and from the handwriting and a few other hints, he deduced that "S. A. Billopp" was a woman, creating a minor furor among his colleagues. Later he was able to confirm his theory when he found Sarah Anne's journal at the Linnaean Society, jumbled among the collections left behind at Burdem Place. The last entry in Sarah Anne's journal was this, most likely copied there soon after she and Mrs. Pearce made their experiments with the swallows:

> *Collinson loaned me one of his books* - An Essay towards the probable Solution of this Question, Whence come the Stork, etc.; or Where those birds do probably make their Recess, etc. *(London, 1703) - with this passage marked for my amusement:*
>
> *"Our migratory birds retire to the moon. They are about two months in retiring thither, and after they are arrived above the lower regions of the air into the thin aether, they will have no occasion for food, as it will not be apt to prey upon the spirits as our lower air. Even on our earth, bears will live upon their fat all the winter; and hence these birds, being very succulent and sanguine, may have their provisions laid up in their bodies for the voyage; or perhaps they are thrown into a state of somnolency by the motion arising from the mutual attraction of the earth and moon."*
>
> *He meant to be kind, I know he did. I cannot bear this situation any longer. Catherine and I are meeting in town to discuss the experiment she's proposed.*

Consequences of Ignoring Gender and Race in Group Work*

Sue V. Rosser

Peter Adams is beginning his third year as an assistant professor in the physics department at GIT (Generic Institute of Technology). He has decided that improving his teaching is a high priority for this academic year. Receiving a grant, although cut by two-thirds from his requested support, means that he can devote less than 100% of his time to thinking about research, at least temporarily.

He had better give some attention to his teaching, since his teaching evaluations have been less than outstanding. His department chair emphasized the importance of teaching in his annual review last year and suggested fairly pointedly that Peter had "room for improvement in that area." Word has been getting around that under the new president, undergraduate teaching is being reaffirmed as a major focus of the institution. Rumor has it that the decision for early promotion and tenure of the best researcher in his department was held up because that person had received poor student and peer evaluations for his teaching. Given the horrendous job market in physics, Peter certainly wanted to hold on to this job.

Peter has decided to focus on improving the instruction in the calculus-based course in introductory physics that he is teaching in the fall term. The course attracts bright students who have indicated their desire to major in physics. Although the department chair always keeps a close watch on that course since it is the feeder course for majors, this year he will monitor it especially carefully because four Howard Hughes Project students—two Hispanics, one male and one female, and two African Americans, one male and one female—have been placed in the course.

In some ways Peter welcomed the pressure that would force him to devote more time to his teaching. The summer had slipped away before he had a chance to study the literature on teaching science, as he intended. But he had ordered and scanned a couple of articles from the massive bibliography that his graduate assistant had gathered on the topic. He also had attended about a third of a half-day symposium at a professional meeting on "Teaching Techniques to Attract Diverse Students in Science." Although not very certain of how they worked, he had picked up the idea that study groups and group projects were what the experts seemed to be talking about as successful ways to attract students, particularly women and minorities, to science. He wished that he had delved more deeply into the literature or had a colleague who was an expert on this with whom he could discuss exactly how to implement these techniques in the classroom, but how difficult could it be anyway? Surely someone who is a nuclear physicist could figure it out. In addition, Peter believed that women and minorities could and should be physicists. Surely that belief and his conviction that he always treats all students the same way, regardless of their gender, will carry them through. After all, aren't being fair and treating all students equally the bottom line in good teaching?

When Peter received his final student roster the first day of class, he was quite pleased. All the students seemed attentive and indeed had the prerequisites required for the course. In addition to 10 white males and six Asian American males (usually representing the extent of diversity in his classes), the class contained two Asian

---

* *Re-Engineering Female Friendly Science*, Sue V. Rosser, Teachers College Press, New York, NY, **1997**, Ch. 2, pp 38-52.

American women, two white women, and the four Howard Hughes students. This diversity was ideal for the new teaching techniques he had decided to try. He divided the students into six study groups for informal support and study outside of class. Each group included one female, one Asian American male, and at least one white male. Four groups had two white males. He placed the African American male in a group with a white woman, and the Hispanic male in a group with an Asian woman.

In the laboratory, he assigned partners to ensure gender and racial equity. The Howard Hughes males were paired with the white females. The four women of color were paired with white males; the six Asian American men were paired with the remaining six white men. Peter used the same laboratory pairs for the "problem competitions" that he introduced during the last 20 minutes of every Friday class. Each week, half of the laboratory pairs went to the blackboard. Peter gave them a problem to solve, which was related to, but different from, the type of problem they had been solving that week. The laboratory partners within a pair competed against each other, in front of the other half of the class, to see who could solve the problem correctly first. The winner of each pair received a piece of candy; the winner for the half of the class received a minor prize, such as a computer disk or a ticket to a basketball game.

During the third week of class, Peter assigned the four groups for the major project, a cooperative effort on which 50% of the course grade depended. Peter announced that all members of a group would receive the same grade on the project. In each group, the leader was to ensure equal participation of all members and smooth functioning of the group. Group one included two white males, two Asian males, a white female, and the Hispanic female as group leader. Group two included three white males, one Asian male, one Asian female, and the black male as leader. Group three included two white males, one as leader, two Asian males, the Hispanic male, and a white female. Group four included three white males, the black female, an Asian female, and one Asian male as leader. Peter was proud to note that he has chosen group leaders to reflect gender and racial equity.

By midterm, when Peter evaluated the effects of his pedagogical changes and group work, he was . . . (Please complete the scenario with your predictions and what happened based upon the situation he has created.)

BACKGROUND: IGNORANCE IS NOT BLISS

Group work has become popular in disciplines ranging from the humanities through the social sciences to the sciences. Faculty in classrooms at levels from preschool through adult education now integrate group work into their teaching. For some, the group work represents an unusual or isolated experience, which they attempt during one class period or for one project or assignment outside of class. For others, group work has become the focal point of the course and has caused them to rethink entirely their curricular content, syllabus organization, problems, laboratory exercises, and methods of teaching.

Although few faculty may be as inexperienced as Peter Adams and create quite the complicated mess likely to result from the scenario depicted above, most faculty have not explored in depth the research surrounding group work. They are relatively unknowledgeable about the actual studies that have examined the circumstances surrounding the general effectiveness of group work—parameters for participant selection, appropriate assignments, grading, and congruence with objectives for the course. Very few have delved into the literature surrounding the significance of gender and race dynamics within groups.

In contrast, a recent study (Brown, 1995) revealed that students, particularly minority women students, understand the significance of group work. Their most highly ranked piece of advice to other women minority students seeking graduate degrees in science, engineering, and mathematics was to become a member of a study group. Many of these minority women revealed that they had been excluded from study groups, which they felt had deterred their success in graduate school.

The lack of attention to the research about group work on the part of science faculty teaching at the college level should not surprise anyone familiar with the world of academic scientists. The considerable demands of research, supervising a laboratory, keeping up with cutting-edge research in their specialty and in any more general areas in which they must teach, coupled with committee assignments within the department and institution as well as those within their broader professional community, leave faculty little time to browse in the literature devoted to pedagogy. Since science faculty holding Ph.D. degrees within a science discipline were trained to do research and probably received little or no formal training in science education or methods of teaching science, most have little acquaintance with the journals publishing research in science education. Although committed to their teaching and desirous of using the most effective methods possible to convey curricular content, few science faculty at the college level are aware of the many parameters that enhance or detract from group work in general or the particular role that gender and race may play in group dynamics.

The purpose of this chapter is to focus on the attention that needs to be paid to gender and race in the group work portion of science teaching. If the dynamics of race and gender are understood and this information is used effectively, group work may enhance learning for all students, especially women and men of color. If these dynamics are ignored or misunderstood, group work actually may inhibit or detract from learning. This chapter will explore parameters such as group size and selection, roles and leadership, assignments and grading, student resistance, and the particular interaction of gender and race with each parameter in group dynamics.

SETTING UP GROUPS

Faculty inexperienced with group work may assume that dividing the class into groups requires little forethought on their part and little explanation to the students. Particularly at the college level, where considerable student autonomy and maturity are assumed and where faculty may have no information about prior achievement or other indicators about students who sign up for their classes, faculty may assume that self-selection and random assignment represent the reasonable alternatives for dividing the class into groups. Simply asking the class to form itself into groups with four members each or to count off by fours often results in situations that inhibit, or fail to enhance, learning for the more vulnerable students. Studies have documented that when students choose their own teams, they tend to choose others whom they know and who are like themselves (Slavin, 1990). Study groups formed outside of class on an informal basis, or lack of supervision by faculty to ensure that all students are included in a group, often leads to exclusion of women and minority students. A graduate student in engineering recounted her difficulties in completing group projects by herself (Brown, 1995) because she found she was excluded from groups formed by white women, as well as by men of color or white men.

One Size Does not Fit All

Ideal group size varies, depending partly on class size and, more important, on the nature of the task or project that the group will carry out. Most research (Johnson & Johnson, 1994) documents that teams should include between three and six members; one tested cooperative learning method for mathematics is called "Groups of Four" (Burns, 1981). Although many faculty report that groups larger than four may encounter problems with smooth functioning, meeting times, and substandard participation by one or more group members, for some tasks larger groups may be useful. Given the current trend toward large cooperative teams for scientific research, some experience with relatively large groups may be appropriate in upper-division undergraduate or graduate level courses.

As with all parameters surrounding group work, the size of the group should be established in light of its objectives or tasks and the reason a team approach best solves this particular problem. For example, for a laboratory experiment in an introductory course that depends on each member of the team having access to the same equipment during a limited period of time, having more than three individuals on the team could be prohibitive. In contrast, for a design project carried out at the senior level over several weeks, where each individual completes a substantial portion of the project alone and then the team synthesizes the results of all team members for a final product or theoretical conclusion (the so-called "jigsaw puzzle" approach), teams composed of five or six members may be needed to complete the function. Setting the group size keeping clearly in mind the objectives of what should be learned from the task and the ideal way to complete it is more likely to lead to the selection of teams of appropriate size.

Selection of Group Members

Faculty selection of groups, as opposed to random or self-selection, permits the faculty member to take race, gender, abilities, and experience into account in determining group membership. An alternative, which is particularly useful in upper-division undergraduate and graduate courses, is to explain the characteristics that need to be considered in group formation to the students and then ask them to use those characteristics to form their own groups. This alternative approach informs the students about the importance of group dynamics and allows them to take some responsibility for forming functional groups. Since learning how to work as a team is one of the objectives of a course in which group work is used, group dynamics and the importance of appropriate group selection are part of the material that the students need to learn, even if the faculty member does the actual group selection.

Science Abilities

What are the significant characteristics that should be taken into account by faculty or students in group selection? A characteristic emphasized in the cooperative learning literature (Oakes, Ormseth, & Camp, 1990) for the K-12 classroom is the importance of forming groups composed of individuals with mixed abilities. In the public schools, where mainstreaming can result in wide variance in student ability within one class and standardized testing requires that all students reach a certain level of proficiency, mixed-ability grouping helps to minimize discrepancies in group progress.

At the college level, the necessity for, or significance of, mixed-ability grouping becomes less evident. Most faculty are forced to assume that students in a particular class have roughly the same abilities. Since most science, engineering, and mathematics courses are lock-step, the fact that students have passed the prerequisite courses diminishes some of the variance based on student ability.

Even in introductory courses, the selective admissions policies of many institutions of higher education ensure relative homogeneity of ability compared with the K-12 mainstreamed classroom. For those institutions with relatively open admissions policies, placement testing for mathematics and science courses for first-year students provides some reduction in variance. Even when considerable variance can be assumed, aside from placement test scores, most college faculty have few indicators of the abilities of their students at the beginning of a course. Because of these factors, the research surrounding the results of mixed-ability grouping generally has little relevance for the college science, engineering, and mathematics classroom. For initial group assignment, most college faculty find it practical not to make ability a major consideration in grouping. After the first examination, project, or other major assessment measure, performance ability might become one of the bases for reassignment of groups (along with other factors such as dysfunctional personality combinations), if it appears that the performance among groups is widely disparate.

Gender and Race: Representation vs. Isolation

In contrast to abilities, race and gender should be considered in initial groupings. Much of the K-12 literature advocates having each group statistically reflect the overall racial/gender mix of the class as a whole (Oakes, Ormseth, & Camp, 1994; Slavin, 1990). This literature supports the notion that having minority perspectives represented in each group will help the majority to understand alternative approaches and ideas contributed by people from diverse backgrounds. For example, a class of 25 students, five African American and 20 white, and approximately equal numbers of males and females, might be divided into five groups, each containing four white students and one African American Student.

Although this apparently "equitable" distribution may be helpful in exposing the majority to minority perspectives, it may in fact be harmful to the minority students. Particularly when the area is a nontraditional major or career choice for the minority, distributing (or isolating) the minority within a majority grouping may lead to the minority dropping out of the group, the course, or the major. The work of Treisman (1992) demonstrated this for African American students at the University of California—Berkeley in mathematics and for Hispanic students at the University of Texas-Austin.

A similar phenomenon may haunt women pursuing nontraditional majors such as science, engineering, and mathematics. Four females and 16 males would not be an uncommon gender distribution in an introductory calculus-based college physics class. Four groups of five individuals each, with one female in each group, appears to represent an equitable (or at least representative) division along gender lines. From the research on study groups at Harvard (Light, 1990) and other work, evidence suggests that women are more likely to drop out of the group if they are the only female, particularly if the subject is a nontraditional one for women, such as science, engineering, or mathematics.

In the college science classroom, where, in contrast to K-12, dropping out or switching majors is a viable option, the research does not support having each group reflect the overall statistical profile of the class with regard to gender and race. Developing individual groups that contain a critical mass (Etzkowitz, Kemelgor, Neuschattz, Uzzi, & Alonzo, 1994), or at least more than one individual of the minority gender or race, leads to less isolation or "spotlighting" of women or men of color. The calculus-based physics class in the example above still might be divided into four groups, with two groups containing two women. Although such groupings result in some groups with only males, the few women in the class are less likely to experience isolation and drop out.

The interaction of race and gender further complicates group selection. While assigning an African American woman to a group whose other members consist only of white men clearly constitutes isolation, questions arise about whether the African American woman is still isolated when another woman (not African American) or an African American male is assigned to the group along with the white males.

Race Reconsidered

Somewhat different, but overlapping, issues surface over definitions of racial minorities for scientific purposes. The National Science Foundation (1994) has defined Asian Americans, because of their relative overrepresentation in science compared with their percentages in the overall population, as not a minority or protected class in science, engineering, and mathematics. Although Asian American men are not a minority, Asian American women are underrepresented. Their gender places Asian American women in a status of underrepresentations similar to that of white women because of their gender. Although the status of Asian American women in science may be similar to that of white women, the differential treatment of Asian American women compared with white women in the overall society means that they will bring a different set of experiences and expectations to the group work.

Defining Asian Americans as not a racially underrepresented group also overlooks diversity among Asian Americans. Although some groups of Asian Americans, such as Chinese Americans and Japanese Americans, may be relatively overrepresented in the SEM workforce, other groups, such as the Hmong, are clearly underrepresented. A growing body of literature (Middlecamp, 1995) has begun to air the misconceptions surrounding the designation of "model minority" or Asian prowess in science, engineering, and mathematics, particularly for some Asian American women, for Asian Americans from particular ethnic backgrounds (such as the Hmong), and for some Asian American men. Considering all Asian Americans as a monolith obscures not only differences in country or origin and language, but also issues such as immigrant or nonimmigrant status and class, which may have more impact on science, engineering, and mathematics representation and performance than does race or gender.

For some groups, race may be a more significant factor than gender for underrepresentation in the sciences. For example, repeated studies (Clewell & Anderson, 1991; Clewell & Ginorio, 1996; Malcolm, Hall, & Brown, 1976) have documented that minority women find both racism and sexism to be significant barriers, with many women ranking racism as more problematic, particularly at the undergraduate level. One study suggested that sexism may be more problematic than racism at the graduate level (Matyas & Malcom, 1991). A complicating factor for minority women that may shed some light on their differential responses was uncovered in the Brown (1995) study. Her study revealed that women pursuing graduate studies in science, engineering, and mathematics who were easily identifiable as nonwhite reported more difficulties with racism; those, such as some Cuban and American Indian women, who stated that others appeared to perceive them as white, reported more problems with sexism.

LEADERSHIP AND OTHER ROLES

Viewing women or all members of a designated racial group as a monolith obscures diversity within groups. Individual differences become especially prominent in the choices of group leaders and assignment of other roles. Assigning a Hispanic woman the role of group leader, assuming that the other group members include a Hispanic male and a white male and female, places her in a role counter to the stereotypes of dominance for her race and gender. Although many individual Hispanic women do not fit the stereotype and feel comfortable with assuming leadership positions, this initial leadership assignment may be difficult for some Hispanic women, perhaps especially in an introductory course when students do not know each other or the faculty member.

For the first group task or project, encouraging students to select their own leader and make their own assignment of roles typically results in group members choosing roles or portions of the task with which they feel most comfortable. This works particularly well when the faculty member has chosen a project where each member has a clearly defined task and where the responsibilities of each role have been delineated. The self-selection of roles within a group facilitates group comfort and functioning, and may mitigate against the loss of autonomy felt due to faculty designation of group membership.

Although initial self-selection of roles may facilitate comfort with group work, roles must be rotated throughout the course, with monitoring to ensure that each student fills each different role for a variety of tasks. Reports from the engineering industry reveal problems when such rotations fail to occur. If students are permitted to stay with the same group all semester, filling essentially the same role to complete several projects or smaller tasks associated with a semester-long project (not an uncommon phenomenon in a senior engineering course), the overall group project may be excellent. All group members may receive an A, because each did her or his part well, particularly since members were permitted to work in their area of strength and the group functioned well as a whole.

Because they were not forced to assume roles with which they had less comfort and for which they held less expertise, however, members may have missed some skills and perspectives developed by the project as a whole, but not available to each member unless rotation of roles occurred. In the industrial setting, where the role available in the work group for the new employee is not the one that she or he assumed in the senior group project, the new employee finds her- or himself at a severe disadvantage.

Gender may be a factor in role preference or comfort. Sometimes a woman prefers or is assigned the role of group manager. Previous socialization as a female in our society may have provided her with better skills in social organization than those of most males. She may be better at interacting with other members of the group to ensure that the group functions smoothly to complete the task on time; her overview of the process as a manager and her writing skills may enhance her abilities to produce the final report on behalf of the entire group. In contrast, a man may have more experience with computers and seem to prefer to interact more with the hardware than with the other people in the group. He may develop superior computer programs to solve design problems encountered by the group and enhance the written portion of the final report with outstanding computer graphics.

As new employees in the industrial setting, both the woman and the man described in the paragraph above may lose their jobs. It is unlikely that the role each assumed repeatedly in the class group will be the only role required for satisfactory performance on the job. Because their group work did not force them to rotate roles, each holds deficiencies in the tasks that each did not have opportunities to develop during the group projects. The man has not learned the social skills and management strategies useful for handling group dynamics; the woman has insufficient experience with hardware and software.

Faculty members must do more than delineate clearly the group roles required for a particular task in order for the group to function smoothly. Monitoring the role assumed by each student during each project permits rotation of roles, while ensuring that individuals will not avoid roles that require new skill development or run counter to gender role expectations.

Group Projects, Assessment, and Considerations of Gender and Race

Group size, selection, and roles may influence gender and race dynamics within groups and overall group functioning in ways not envisioned by Peter Adams and other faculty who have not read the literature or had experience with group work in the science classroom. In a similar fashion, inexperienced faculty may fail to consider the changes in types of assignments, projects, or methods of assessment necessitated by group work, or how best to mesh group work with overall course objectives.

Group-Appropriate Assignments

An error made by some faculty attempting group work in the science, engineering, or mathematics classroom is to make no change in the assignments traditionally used in the course; they simply expect the students to do the same assignments in groups. If the problems, experiments, or projects traditionally used in the course have been designed for individual work, they are unlikely to be particularly appropriate for group work. Students rapidly perceive that this is a task they could complete on their own, perhaps even more quickly, and resent the group work as an unnecessary bother than complicates their schedule, especially if the group work is expected to take place outside of class.

In the work world, teams or groups are used to solve problems where the perspective, skills, and knowledge of one individual are inadequate or where the complexity and/or amount of time required to solve the project are excessive for one person. Individuals actively depend on the other members of the group because they understand that they cannot complete the project alone. They cooperate with the group because their self-interest is dependent on successful completion of the team project.

Problems or projects assigned to groups in the science, engineering, or mathematics classroom also need to reflect complexity, time pressures, and a variety of skills to reveal that team work is critical to their solution. For the faculty member who previously has taught the course using traditional, individual assignments, the transition to groups involves substantial rethinking and work. Contrary to an expectation held by some faculty that group work will lighten their class preparation, it involves considerably more time and preparation, especially the first few times the course is taught using groups.

Development of guided design problems, complex projects, and multi-faceted experiments with appropriate and clearly defined roles (Brown & Campione, 1994) usually requires considerable upfront time for faculty. It may take more time than conceptualizing a new course or redesigning the content of a previously taught course while retaining the traditional teaching methods. A faculty member who has never worked in a setting where teams are used or taken a course where group work figured extensively will be rethinking his or her own ingrained patterns of teaching and problem solving, while trying to develop assignments appropriate for group work.

Problems and projects developed need to draw on and reflect experiences and issues significant for both males and females and individuals of all races. Overall course objectives and tailoring of problems for the team structure serve as the primary principles guiding this development. However, gender or race may influence the perception of the faculty member if she or he has selected to undertake group work because of the research suggesting that women and men of color may respond particularly well to cooperative learning and practical applications. When that becomes a major motivation driving group work, the faculty member may come to wonder whether it is really worth all the extra work to redesign the course, especially if the students demonstrate an initial response to group work that is less than enthusiastic.

One source of student resistance may result from having assignments that are inappropriate for group work. Because until recently most students have experienced more passive forms of content delivery, such as lectures, in science, engineering, and mathematics courses, they do not expect the more active form of learning required by group work. This expectation for passive learning leads most students to resist group work initially, especially because it requires more work for them.

Some of the most resistant students often are high-ability, high-achieving minority women. When many faculty using group work reported this to me (Rosser & Kelly, 1994), I explored some of the reasons for the resistance. Although no statistical data were collected about this, the belief of the faculty, based on conversations with these students, was that the resistance stemmed from the fact that the students had worked very long and hard to develop skills that helped them to do well in science, engineering, and mathematics. Many of them also had significant outside responsibilities such as families and full-time employment. They viewed group work as a situation in which they again would be asked to be responsible for someone else (i.e., they would have to teach the other students and bring them along) and/or as an additional complication to an already difficult schedule. After the initial resistance, almost all students, including these high-achieving minority females, reported high satisfaction with group work as an experience that aided their learning, if assignments were appropriate for team efforts.

Methods of Assessment

Concerns over methods by which group work will be assessed and reflected in the final individual course grade often contribute to student resistance. Many faculty also express discomfort with assessment of group work. Studies (Champagne & Newell, 1992) reveal a range of approaches from not assessing or counting group work at all to make the common group grade count as the individual final grade for each student. Faculty also vary in the extent to which they take complete responsibility for grade assignment or consider student assessments of individual contributions to the group effort.

The sample peer evaluation form reproduced in Figure 1 provides one example of how a faculty member using team projects in a large course assesses the group work.

Figure 1. Peer Evaluation Form

Objective: I will assign a score to each team project and then adjust each member's individual score by his or her evaluation from peers. These evaluations provide you with protection against team members who wish to receive a good grade without doing the work.

Procedure: You are to assign 100 points among yourself and the other members of the group. If, say, there are four members in your group and all made equal contributions, then each member, including yourself, would receive 25 points. If, however, three members did most of the work and the fourth person malingered, your point assignment might be 27 points to each of the three workers and only 19 points to the malinger. (Note: Each team member must submit a PE form on April 23 when the project is due.)

Print or clearly write names of all group members in the spaces provided:
How would you characterize the amount of *time and effort* spent and the *overall* contribution of each group member?

Member 1(yourself) ______________________ points __________
Member 2 ______________________________ points __________
Member 3 ______________________________ points __________
Member 4 ______________________________ points __________
Member 5 ______________________________ points __________
TOTAL __________

In the following space and on the back of the page, explain why you have given a team member (or members) fewer points that you have given yourself. *Note*: All team members will receive equal credit unless a written explanation is provided. Your explanation will be held in confidence. I will share the explanation with the person receiving the reduced grade but I will *not* identify you by name.

In this example, the faculty member assigns an overall project grade and adjusts the individual student grades based on students' assessment of their own work and that of the other team members. This assessment conveys to students that the group work is important (i.e., it counts as part of the course grade), that the group product as a whole will be assessed (i.e., the faculty member will grade the product, possibly in relationship to other group products or against a standard of what is a correct answer or good outcome), and that the perceptions of the teammates as to who is doing the work also count.

In cases where students undertake group work, but the grade assigned is based on a laboratory write-up done individually by each student, a different message is conveyed. Students will recognize the significance of individual performance. Group work and cooperative learning do not count, except insofar as they enhance the individual laboratory report.

In contrast, making the group work count for 100% of the course grade and having the group grade be the same for all individuals in the group relay a very different message. This assessment implies that group work counts for everything and that individuals will do well only when the group functions to produce an excellent product, which in turn will benefit the grade of each group member.

This last example may seem to give extreme emphasis to group work and penalize individuals in groups with one or more nonfunctioning members; in certain instances, this may be an appropriate assessment. In a senior-level engineering course, after which most students will take jobs in industry, a major course objective may be to teach students to function well in teams. In industry, all members of a team may experience small salary increases or lose their jobs, if they fail repeatedly to function well and produce. Perhaps the best method to prepare students to grasp the importance of group functioning in the work world is to have all members of a team receive the same individual grade.

This last example stresses the important principles that should underpin all assessments of group work: the assessment should be appropriate for the significance accorded group work and consonant with the course objectives for the use of group work. Students receive conflicting messages when teaching students to work in teams is listed as a major course objective, but group work does not count toward the final grade. It appears equally inappropriate to base the entire final grade of each student on the group project grade, while not including group work as a primary objective of the course.

Issues of race and gender may enter into assessment of group work, particularly during the stage of evaluation of the contributions of individuals to the group. Research on gender roles in group dynamics documents differences in the ways individuals are perceived and treated, based on their gender. Studies in formal groups containing both men and women have demonstrated that men talk more than women and exert more control over the topic of conversation (Kramarae, 1980; Tannen, 1990). White men also interrupt women much more frequently than women interrupt men, and their interruptions more often introduce trivial comments or statements that end or change the focus of the women's discussion (Zimmerman & West, 1975). Further, the more frequent use by women of "tag" questions ("This is true, don't you think?"), excessive use of qualifiers, and excessively polite and deferential speech forms may make women's comments more easily ignored or seem to carry less weight in a group. Another form of ignoring women's contributions occurs when an idea or suggestion initially made by a woman is attributed later by another group member to a male in the group, who then receives credit by having his name attached to the idea.

In peer evaluations of individual group members by themselves and others in the group, the contributions of women may be undervalued because of this gender bias that operates against females in group interactions. In science, engineering, and mathematics groups, the contributions of females may be further underrated if they have played the role of group manager or facilitator. Particularly if they have done an excellent job of making all members feel that they are part of the group and of facilitating smooth group functioning, much of their work may have been "behind the scenes" or barely perceptible to other group members. In the sciences, where experimental data, ability to work with hardware or equipment, and theorizing may be highly valued, individuals in those roles may be perceived as particularly valuable or hardworking, which may be reflected in the peer evaluations they receive from others and themselves. In contrast, the role of group manager or facilitator may be devalued and receive lower peer evaluations, particularly when that role is consonant with the stereotypical expectations for women.

CONCLUSIONS

Most faculty who undertake group work in the science classroom assume that it represents a positive force for retaining women and men of color, if they consider the impact of group work on those students at all. Aware that cooperative learning and reduction of competition may attract women and men from some ethnic/cultural backgrounds, faculty assume that use of a group format will facilitate learning for individuals traditionally underrepresented in science, engineering, and mathematics. Without some knowledge of the role that gender and race dynamics may play in group interactions, groups may be set up to function in ways that deter men of color and women. In contrast, when faculty give attention to gender and race in determining group size and composition, selection of leaders and other roles for individuals within the group, and particular assignments and grading, they may enhance the learning for all students, especially men of color and women.

REFERENCES

Brown, A. L. & Campione, J. C. (1994). Guided discovery in a community of learners. In K. McGilly (Ed.), *Classroom lessons: integrating cognitive theory and classroom practice* (pp. 229-270). Cambridge, MA: MIT Press.

Brown, S. V. (1995). Minority women in graduate science and engineering education [Abstract]. In *Unity in diversity* (p. 79). Atlanta, GA: AAAS Annual Meeting and Science Innovation Exposition.

Burns, M. (1981, September). Groups of four: Solving the management problem. *Learning*, pp. 46-51.

Champagne, A. B. & Newell, S. T. (1992). Directions for research and development: Alternative methods of assessing scientific literacy. *Journal of Research in Science Teaching*, *29(8)*, 841-860.

Clewell, B. C. & Anderson, B. T. (1991). Women of color in mathematics, science and engineering: A review of the literature. Washington, DC: Center for Women Policy Studies.

Clewell, B. C. & Ginorio, A. B. (1996). Examining women's progress in the sciences from the perspective of diversity. In C.-S. Davis, A. B. Ginorio, C. S. Hollenshead, B. B. Lazarus, P. M. Rayman, and Associates (Eds.). *The equity equation* (pp. 163-231). San Francisco: Jossey-Bass.

Etzkowitz, H., Kemelgor, C., Neuschattz, M., Uzzi, B., & Alonzo, J. (1994). The paradox of critical mass of women in science. *Science*, *266*, 51-54.

Johnson, D. W., & Johnson, F. (1994). *Joining together: Group theory and group skills* (5th ed.). Boston: Allyn & Bacon.

Kramarae, C. (Ed.) (1980). *The voices and words of women and men.* London: Pergamon Press.

Light, R. (1990). *Explorations with students and faculty about teaching, learning, and student life*. Cambridge, MA: Harvard University Press.

Malcon, S. M., Hall, P. Q., & Brown, J. W. (1976). *The double-bind: The price of being a minority woman in science.* Washington, DC: American Association for the Advancement of Science.

Matyas, M. & Malcom, S. M. (1991). *Investing in human potential: science and engineering at the crossroads.* Washington, DC: American Association for the Advancement of Science.

Middlecamp, C. (1995). Culturally inclusive chemistry: In S. V. Rosser (Ed.), *Teaching the majority* (pp. 79-97). New York: Teachers College Press.

National Science Foundation (1994). *Women, minorities, and persons with disabilities in science and engineering*. Arlington, VA: National Science Foundation.

Oakes, J. (1990). *Lost talent: The under participation of women, minorities, and disabled persons in science*. Washington, DC: National Science Foundation, with the Rand Corporation.

Rosser, S. V. & Kelly, B. (1994). *Educating women for success in science and mathematics* (NSF Project HRD 9053892). West Columbia: University of South Carolina Publications.

Slavin, R. E. (1990). *Cooperative learning, theory, research, and practice*. Englewood Cliffs, NJ: Prentice-Hall.

Tannen, D. (1990). *You just don't understand*. New York: Ballantine.

Treisman, P. U. (1992). Studying students studying calculus: A look at the lives of minority mathematics students in college. *The College Mathematics Journal, 23(5)*, 362-372.

Zimmerman, D. H. & West, C. (1975). Sex roles, interruptions and silences in conversation. In B. Thorne & N. Henley (Eds.), *Language and sex: Differences and dominance* (pp. 105-129). Rowley, MA: Newbury House.

Addressing Homophobia, Biphobia, and Heterosexism in Workshop

Francisco Ramirez, workshop leader

The organic chemistry workshop is a social and educational environment that promotes active learning. The constant interaction between the students and the leader has a positive impact on the learning, but due to the cultural diversity of the group, some conflicts may also arise. For instance, homophobic, biphobic, and heterosexist attitudes have a negative impact in the development and learning of gay, lesbian, and bisexual students.

In a society that is intolerant toward homosexuality and bisexuality, like ours, homophobia, biphobia, and heterosexism are common. Antigay messages are ubiquitous. The media, political and religious organizations, and even schools discriminate and oppress gay, lesbian, and bisexual individuals. Thus, it is not surprising that by the time students get to college, many of them have deeply rooted homophobic, biphobic, and heterosexist views. Many of these negative attitudes are unconscious due to the pervasiveness of negative stereotypes. The lack of positive roles makes challenging these stereotypes a difficult task.

Many gays, lesbians, and bisexuals begin the process of self-acceptance and disclosure during the college years. During this time, the attitudes they encounter from others influence their psychosexual development greatly, which can (and in fact often does) manifest itself in their academic performance. Ignoring the issues of homophobia, biphobia, and heterosexism in workshops means discriminating against an entire group of students. When gay, lesbian, and bisexual students encounter these oppressing attitudes in the workshop, they feel ignored and rejected, which, in turn, obstructs their learning experience.

There are many ways in which the leader can modify his or her workshop to be entirely welcoming to students of all sexual orientations. For instance:

- The workshop leaders should address their own attitudes toward homosexuality and bisexuality. Reading books on the subject, or talking to a gay, lesbian, or bisexual friend are ways in which they can develop a better understanding of the lives of gays, lesbians, and bisexuals.

- Workshop leaders should familiarize themselves with statements that are homophobic, biphobic, or heterosexist, and refrain from making them during workshop. For example, if a student mentions something about a date or a loved one, it should not be assumed that his or her significant other is of the opposite sex.

- Defamatory humor or other forms of negative comments should be discouraged during the workshop. For example, if a student begins telling a “queer” joke, the leader should state that such humor is not tolerated in the workshop.

William Skawinski[*]

by Michael Woods

The computer terminal beeps to signal an incoming e-mail message, and William J. Skawinski saves the draft of a scientific report, closes out the word processing program, takes the message, and responds. Before calling up the report again, he takes a few moments to note the position of atoms in a computer-graphics-generated model of a cyclodextrin molecule. Skawinski and a group of colleagues at New Jersey Institute of Technology (NJIT) in Newark are studying the compounds, which are related to a diuretic drug. He also shuffles through a sheaf of nuclear magnetic resonance (NMR) spectra.

"I've got three reports and a grant application in different stages of completion," Skawinski pointed out. "I've got to get those done. I'm teaching an organic chemistry course to sophomores and juniors in the autumn, and the lesson plan needs work. I've got journals and this week's *Chemical & Engineering News* to read. Where do I find the time?"

These, of course, are activities and concerns common to many chemists. They may seem a bit unusual in this instance because Bill Skawinski is blind.

"Blindness is much less of a problem in science today than most people think," he said. "Computers, the Internet, and CD-ROM are among the best imaginable developments for the visually impaired scientist. They give quick, accurate, and inexpensive access to all kinds of information that once was beyond reach."

People with disabilities as a group are seriously underrepresented in science and engineering professions. But studies suggest that visually impaired scientists are among the most underrepresented. There may be fewer than 100 blind men and women employed in chemistry, for instance. Research has revealed the role of negative attitudes and lack of information about the capabilities of people with disabilities in discouraging careers in science. "Some of the negative attitudes may arise from a lack of information about the technology available to help blind and visually impaired scientists, students, and teachers access and use information," said Skawinski.

In his research and teaching, Skawinski uses technology that many sighted chemists don't even know exists. These include high-tech devices, like a "talking computer," which Skawinski regards as the single greatest advance in giving visually impaired people access to information.

But they also include very ordinary materials. Skawinski is part of a research group, headed by Carol A. Venanzi, that uses computational techniques and NMR in studying how drugs interact with sodium channels in the cell membrane. How does a blind scientist "see" NMR spectra? For relatively simple spectra, without a lot of fine detail, all it takes is the adhesive glitter paint that kids use in school craft and home art projects. A sighted person traces the NMR plots with glitter paint, which leaves a raised

---

[*] Working Chemists with Disabilities, Expanding Opportunities in Science," Co-Editors: T. A. Blumenkopf, V. Stern, A. B. Swanson, and H. D. Wohlers, Written by Michael Woods, Published by the American Chemical Society Committee on Chemists with Disabilities, Washington, D.C., **1996**, pp 56-61.

line. Skawinski then simply feels and "reads" the plot with his fingers. By stacking several, Skawinski can compare differences by feeling the offset in the lines.

"It works, and it works extremely well," says Skawinski. "Sometimes you can get caught up in the technology, and think that everything takes a high-tech solution. But occasionally, the most effective answers can be simple and inexpensive."

Bill Skawinski began losing his vision to retinitis pigmentosa while still in first grade. Retinitis pigmentosa is a degenerative condition of the retina, the light-sensitive tissue at the back of the eye. The disease first attacks the retina's rod cells, which sense various intensities of light. People with the condition first notice diminished night vision and develop "night blindness." Then the disease turns to the retina's cone cells, which are sensitive to color. Daytime vision begins to deteriorate as well. Strangely, the deterioration occurs from the outer circumference of the visual field and moves inward. The field of vision is slowly reduced, in a telescoping effect, to a smaller and smaller central core. Then, nothing. Retinitis pigmentosa's cause is unknown, and there is no treatment. By eighth grade, Bill couldn't see the blackboard in school. By college, he could barely read with a magnifying lens. Four years after getting his B.S. degree in chemistry from the Stevens Institute of Technology, he had no vision at all.

"As the vision got worse in undergraduate school, I never seriously entertained the idea of changing to another major," Skawinski said. "I grew up loving science. All my toys as a kid involved science. There was just no other conceivable career." Bill's mom and dad, though they always supported his interest in science, thought that he might be too young for the Gilbert chemistry set that he wanted at age nine. So, guided by a book he obtained from the local library, Bill assembled his own elaborate chemistry set from household chemicals and glassware from a laboratory supply house. His high school rocket club experimented with high-energy propellants. One of their rockets reached an altitude of 42,000 feet.

Skawinski went into a rehabilitation program after getting his undergraduate degree. He learned Braille, how to navigate with a cane, and other adaptations. He went on to earn an M.S. degree in chemistry from NJIT and a Ph.D. in chemistry from Rutgers University. He relied on Braille, molecular model kits, recorded books, readers, and other assistive technologies then available for the visually impaired.

The basic accommodation for a visually impaired scientist is already a fixture in most workplaces. It is a personal computer, equipped with accessory hardware and software. In Bill Skawinski's instance, it was the only major physical accommodation needed for his job. Skawinski's talking computer is an ordinary desktop computer outfitted with a voice-synthesizer card and screen-reader software. The voice repeats each keystroke through a speaker: "T-h-e/space/ v-o-i-c-e/space/r-e-p-e-a-t-s/space/e-a-c-h/space/k-e-y/space/s-t-r-o-k-e/space/t-h-r-o-u-g-h/space/a/s-p-e-a-k-e-r/period/" Though Skawinski does sometimes use the synthesizer to echo each keystroke, for most writing, he sets it to read back complete words only. A user can also adjust the speed of vocalization so that keyboarding can be very rapid. Skawinski uses a standard keyboard. He is a fast typist, who needs only the two tiny elevated dots on the "F" and "J" keys as landmarks.

A user can also scroll across the screen to get an audible reading of the content. This might be a list of files in a directory or a list of messages in an e-mail inbox. When booted-up for the first time, the machine vocalizes the usual routine that sighted users observe on the screen: It asks "User name?" "Password?" and observes "You have six e-mail messages." If requested, it then reads the messages. Skawinski routinely uses it to access a supercomputer and for many standard scientific applications. Both the card and the software are widely available. Skawinski's home and laptop computers are equipped in a similar fashion.

Listening to and remembering the audible computer output from spreadsheets and other lengthy tabular data can be difficult. So Skawinski uses an 80-column Braille display. It outputs digital tabular data into tactical forms as rows of raised metal Braille prongs on a panel. By running his fingers along the panel, Skawinski can read, re-read, and compare columns of experimental data. For visually impaired individuals like Skawinski who read Braille, printers that output with raised dots are available.

Scanners have been another advance in accessing information. Scanners transform text and graphics into computer files that can be edited with word processing programs. Scanners come as a standard feature on some new personal computers and are available as add-ons for a few hundred dollars. They allow visually impaired people to rapidly input login documents and access them with the computer's speech synthesizer. Most programs for computer fax/modems have optical character recognition (OCR) software. OCR converts faxes into editable digital files that a blind person's computer can output through the voice synthesizer.

Like many other visually impaired people, Skawinski is an enthusiastic user of the large assortment of recorded materials, including books and magazines. The textbook for his organic chemistry course is on audiotape. A Braille version of the text would occupy several feet of shelf space. The tapes and a teaching assistant (TA) are all the accommodations necessary for the course. The TA will grade exams (generated from software provided by the textbook publisher) and help with other tasks such as preparing transparencies for the overhead projector that Skawinski uses for lectures. What about drawing structures and explaining reactions on the chalkboard? Skawinski makes students active participants in lecture sections by giving them a chance to write and diagram on the chalkboard. The approach enhances students' learning while accommodating Skawinski's disability.

The Internet, and its most popular component, the World Wide Web (WWW), are giving visually impaired people greater access to scientific literature. A growing number of journals have online versions of their paper-and-ink editions, as part of a trend toward cutting publication costs and speeding publication of new research findings. Libraries likewise are turning to CD-ROMs to reduce the need for storage space. Skawinski uses the American Chemical Society (ACS) Web site (http://www.acs.org) for access to the *Journal of Physical Chemistry* and other ACS services, and the Chemical Abstracts Service's STN for access to the flagship *Journal of the American Chemical Society* (JACS) and other journals. Skawinski reads *Chemical & Engineering News* each week from the ACS Web site, or via e-mail. JACS and other ACS publications also are available on CD-ROMs. Web pages of other scientific organizations, such as the American Association for the Advancement of Science (AAAS), provide access to advances in chemistry and other fields. The AAAS Web site (http://www.aaas.org), for instance, carries articles from current editions of *Science*.

For back issues of journals, and current issues of journals that are not yet online, Skawinski uses volunteer or hired "readers." Some visually impaired people prefer content read directly to them. Others, including Skawinski, prefer having text read into a tape recorder. A tape provides a permanent record that Skawinski can replay whenever necessary.

Years ago, Skawinski and mechanical engineering professor Ira Cochin from NJIT developed "talking" and tactile laboratory instruments. Cochin, one of several people who Skawinski regards as a mentor, offered invaluable help and encouragement. The instruments include a laboratory balance and a spectroscope with audible output. The device emits a tone that varies in pitch according to the height of the spectral line. Most modern laboratory instruments have digital output and computer interfaces. So it is possible to convert the output into a form accessible to the visually impaired.

Many employers may imagine a visually impaired chemist in the laboratory as an accident waiting to happen. Skawinski notes, however, that visual impairment imposes an extraordinary sense of orderliness and neatness on an individual. Neatness and orderliness also are cardinal rules for safety. Like most people with disabilities, those with limited vision learn to become excellent managers, keeping objects in order so they can easily be found. Skawinski recalls that as an undergraduate and graduate student, his lab bench was the neatest. Accidents happened, but to the sighted students trying to work in cluttered spaces, not to Bill. During a renovation project at NJIT, he walked with assurance down corridors temporarily cluttered with desks, chairs, and other items. Companions say that Skawinski is equally surefooted on hiking expeditions through the backwoods.

It is almost axiomatic that accommodations developed for people with disabilities sometimes become valuable for others. One of the classic illustrations emerged from the work of Skawinski and Carol Venanzi, who is a professor of chemistry at NJIT.

Many scientific disciplines rely heavily on three-dimensional (3-D) computer graphics to display complex data. In chemistry, for instance, 3-D molecular models can depict the shape of an enzyme's active site or of a cell-surface receptor site. From studying the image, chemists may get clues about how to make improved enzymes that can be used for industrial processes or drugs that block a cell-surface receptor and prevent a disease.

However, 3-D images on a computer screen are practically useless for blind and visually disabled scientists and students. For years, Skawinski thought about making physical models of the video images that could be read, almost like Braille. The idea moved toward reality when NJIT's Center for Manufacturing Systems purchased a laser stereolithography apparatus. Laser stereolithography is a rapid prototyping process that produces plastic models from an image created by computer-assisted design (CAD) software. Under computer guidance, a laser beam traces the shape of an object in a container of photosensitive resin. It polymerizes the resin, creating a plastic model of an image layer by layer from bottom to top. Stereolithography is widely used by the automotive, aerospace, and other industries to make prototypes of components.

Skawinski recognized its potential for making molecular models, and with Venanzi got a $360,000 National Science Foundation grant to investigate production of models of molecular structures and other objects for visually impaired scientists and students. Each atom in a model has a characteristic texture, so Skawinski can quickly distinguish nitrogen from oxygen, carbon, and hydrogen. The models have many advantages over traditional stick-type molecular models. The technique can be applied to virtually any image from chemistry, physics, and biology so long as the image can be described mathematically in a CAD program.

Sighted students and researchers immediately became the biggest fans of the models. It turned out they, too, had difficulty understanding the 3-D spatial relationship displayed on a computer screen. The ability to hold the molecule, to touch and feel the bond angles and relative sizes of the atoms, lent a new dimension to their understanding.

"It is important to remember that not every piece of equipment is necessary for every visually impaired person," Skawinski said. "What an individual needs depends on the kind and severity of the impairment, their job responsibilities, and personal preference." Not every adaptation need be high tech.

Computers and modems have made it possible for some employees to do much of their normal work at home. Skawinski said that NJIT has been accommodating in his work schedule, so that he need not make the 14-mile commute to campus from Garfield, New Jersey, every day. He usually comes in for meetings, seminars, and teaching, and telecommutes on other days. The NJIT administration has been cooperative, yes. But Skawinski says administration and staff have gone far beyond that and have encouraged and supported him.

Other chemists at NJIT say they have had to make few accommodations in working with a blind colleague. When preparing a research paper, for instance, co-workers usually concentrate more on the graphics and figures, while Skawinski works more on the text. "There's really been no adjustment or minimal adjustment on our part," Venanzi said. "He's a productive, valued colleague, and the blindness is almost incidental." Skawinski realized how much attitudes had changed when one day a colleague described the size of a plastic molecular model by holding two of his fingers apart and telling him, "It's about this big." Skawinski believes that his accommodation—a change in attitudes and preconceptions about scientists with disabilities—may be the most important of all.

As Bill Skawinski's life and career demonstrate, it is very possible, indeed, to find talent, productivity, and creativity in solving problems and disability together in the same employee.

Some Common Myths about Disabilities

Vicki Roth

Here are some common myths about the nature of disabilities, followed by alternative views.

- *Myth 1*

A "learning disability" is just a polite way to refer to lower overall intelligence and abilities.

Learning disabilities are quite different from global impairments. In fact, inherent within the definition of learning disabilities is a discrepancy between demonstrated intelligence and specific functioning. It is possible for a student to be both gifted and learning disabled at the same time; many students with LD attending college belong in this category.

- *Myth 2*

Given the proper instruction, students can grow out of their learning disabilities.

Learning disabled students can and do acquire improved skills, but the learning disabilities themselves are not cured in the process. Because learning disabilities are now thought to be lifelong conditions, it is more appropriate to talk about "compensation" than a "cure."

- *Myth 3*

Students with disabilities have "attitude problems."

It is certainly true that the frustrations of living with a disability can cause a loss of self-esteem, and this in turn can look like "a bad attitude." But this is a result and not a cause of a disability. It is critical that faculty and staff understand the level of dedication many students with disabilities bring to the college. They enroll knowing that, in many cases, they will need to put in double the effort of students without these limitations. A little understanding can go a long way to help these students retain the level of confidence and motivation they will need to succeed.

- *Myth 4*

Accommodating the needs of a student with disabilities means watering down course requirements.

Teaching a student with special learning needs does not mean "less." It may, however, mean "different." The goal posts don't change position, but the way in which the student runs down the field may differ. In other words, the instructional goal should be to find ways to work around the area of deficit in order to impart and to evaluate the same body of information and sets of skills.

- *Myth 5*

Students use disabilities as an excuse to get out of work.

Students with disabilities may need help from their professors and TAs to see the essential information in a course. Their purpose is not to avoid work, but rather to focus their efforts. Many students with disabilities routinely invest far more time in their studies than their non-disabled peers.

- *Myth 6*

Accommodations for disabilities give students an unfair advantage.

Without modifications, common forms of instruction and examination often inadvertently reflect a student's disability rather than the subject at hand. For instance, a student with dyslexia may perform poorly on an essay exam question, even with a large fund of knowledge on the topic, because the dyslexia makes on-the-spot writing very difficult. An alternative examination may factor out the dyslexia and allow the instructor to measure the student's understanding of the subject instead of the degree of disability. So good accommodations aren't unfair; they simply level the playing field.

APPENDIX

Troubleshooting

Even with the best preparation, something in your workshop experience may throw you a curve. This section is intended to help you find solutions to some of things that can go a bit haywire.

The Shy Student

If you have a quiet student in your group, remember that she or he may be gaining more from the group experience than is apparent at first. It may just take this student a longer time to feel comfortable with the group. While she or he is warming up, there is much you can do to help, like keeping the quiet and steady encouragement coming, setting up the paired problem solving technique described in Chapter Five, and arranging your session so that the shy student must offer or request information from other participants in order to complete the task.

After a couple of sessions, if the student is still not actively engaged in the group, assess this person's skill level. Is she or he bored, behind the group in math or other skills, or unsure about procedures? Another group member could be encouraged to step in to help. Or it may be necessary for the leader to offer the assistance needed to allow the student to proceed through the workshops. If you detect skill or knowledge deficits that will hinder the student's success, notify the professor or the program director. Generally, the student who is shy becomes a fully participating member once she or he is integrated into the group.

The Domineering Student

The domineering student may be doing all the work, refusing to let others talk, ordering people around, or making decisions without first obtaining group consensus. This student needs to become a cooperating member and to allow the rest of the students their own space in the workshop activities.

You can redirect the group by structuring materials so that the task can't be completed unless everyone participates. You will find it beneficial to encourage and listen to each student's input while not allowing any interruptions. You will need to ensure that the leadership roles rotate throughout the group, and that all members respect each other's right to contribute. Finally, make use of the domineering student's outgoing approach to life. Put him or her to work; ask for assistance in helping the other students work their way through the problems.

If none of these strategies works, take the student aside at the beginning or end of the session for a private conversation on the matter. State that you are concerned about the other students' opportunity to make contributions to the group and ask for help. This almost always works.

The Group That Won't Talk or Won't Stay on Task

Each group develops a personality of its own. However, if a group is still not talking much after the first few sessions, the leader needs to take action.

The leader should look at the seating arrangements; can the participants see and hear one another easily? What time does the workshop meet? If the students are famished, ordering in pizza or encouraging them to bring a snack may give them energy and facilitate the process of getting acquainted and building a working relationship. If the room is too warm, the students may be sleepy.

Some groups talk about everything but the subject at hand and fail to complete workshop activities. If you have a workshop like this, look at the composition of your group to see which members are contributing to the problem. Would a change of roles give the most troublesome students the motivation to do the work? Are the students discouraged because the workshop problems are too hard or too easy? Do they have the social skills necessary to work together as a group?

Perhaps the students are unclear about the benefits of the workshop. If they see the workshop as just one more thing the instructor is making them do, they will not understand that the workshop exists to ensure their own success, nor will they have a sense of the interdependence needed to keep a group working.

For a group like this, it might be helpful to establish a set of expectations for the workshop. These could include:

- Taking responsibility for their learning and the learning of others.
- Providing assistance to others.
- Accepting other members.
- Reaching outcomes through consensus.
- Respecting each member's attempts to learn the material.

Help them see that failing to do work can negatively affect the learning—and therefore the grades—of *all* students in the group. This way, the more motivated participants are likely to see that this outcome would be unacceptable, so they'll put pressure on the others to stay involved.

If the group remains disengaged, ask them what they need to be more productive. They may come up with a workable solution to their problem. If they don't, invite an outside person, like another workshop leader, to spend some time with the group. Often an impartial observer can see where the problem lies and make suggestions for growth in the way the group functions.

The Overcommitted Leader

In the last section, we discussed the importance of setting boundaries. It is also vital for leaders to understand that their role as workshop facilitators requires dutiful attention to the group, which includes thorough preparation. But we do not want you to hurl yourself into this role so intensely that you lose sight of the rest your courses, your friends and family, your job, and the rest and relaxation you need.

If you find yourself overloaded, the best thing for you to do is to admit you could use a hand in getting your priorities realigned. Leaders with balanced lives are more effective in the workshop. It may be helpful for you to talk with the workshop program director for help in finding the right balance. Don't forget your campus learning center and counseling office; they help students work through these issues every day. They are likely to have some good strategies, not to mention a friendly ear.

The Student with Personal Issues

Our workshop participants are students in other courses, they are friends and members of families, they may be employees. Sometimes they unburden themselves to their workshop leaders. Being the kind person you are, it may be tempting for you to take on responsibility for their problems, but you will be most helpful by listening carefully, and then deciding if the issues merit referral to other resources, like the counseling office for a personal or family matter, or to the study skills center for study strategy and time management difficulties, or to faculty advisers. It helps to complete the list at the end of this section so you will have names, campus locations, and phone numbers of offices you might need to refer students to.

A very common "emergency" is their panic about an upcoming test in the workshop course. It is important for you to inform your students ahead of time about your availability for last-minute questions and tutorials. If you agree to offer additional assistance, be clear about the hours you will accept phone calls and how much help you are prepared to offer. Encourage them also to seek help from the instructor and the TAs.

Setting boundaries is important for peer leaders and their students. They will gain much more from the workshop if they understand and respect these limits, and you will be better able to focus your energies on helping your students in the workshop sessions.

All of the above are general suggestions for managing troublesome situations so that they don't interfere with workshop goals and objectives. You will find that some problems resolve themselves and some improve with assistance from you. The basic rule is to seek assistance before things get out of hand or you become overwhelmed. Always remember that your program director and the faculty in the discipline are available to help you.

Some Helpful Numbers

Take a few minutes to collect the names and numbers for the following resources on your campus. Chances are good that some of your workshop participants will need help from some of these offices during the course of the term. If you can provide detailed contact information to your students, your students are more likely to follow through.

name	campus address	phone	email
academic advising			
counseling office			
course instructor			
dean of students			
financial aid			
health service			
residential life			

Self-Rating Checklist

Please rate your leadership and your workshop sessions according to the questions below. Feel free to make additional comments and observations at any point. Note that you are not being asked to put your names on this form, so please be candid in your replies.

		Very seldom	Sometimes	Often	Almost always
1.	I understand the goals of the workshop program.	1	2	3	4
2.	I understand the professor's goals for the modules and for the course as a whole.	1	2	3	4
3.	I prepare well for the workshops.	1	2	3	4
4.	I can get the workshop sessions started easily.	1	2	3	4
5.	I am happy with the amount of control I have in the workshop (not too much, not too little).	1	2	3	4
6.	I am patient.	1	2	3	4
7.	All of the students participate on a regular basis.	1	2	3	4
8.	I am adept at keeping the conversation going among the students.	1	2	3	4
9.	I am good at asking questions that help students approach the problems.	1	2	3	4
10.	My students look to me for help the appropriate amount of time (not too often, not too seldom)	1	2	3	4
11.	The workshop students talk easily with each other.	1	2	3	4
12.	I am able to tell people they are incorrect in a constructive way.	1	2	3	4
13.	I am confident about explaining the material when it is appropriate for me to do so.	1	2	3	4

14. I know how to handle it when someone asks me a question I can't answer.	1	2	3	4
15. My workshop stays on task the right amount of time each week.	1	2	3	4
16. I can get the students back on track when they get distracted.	1	2	3	4
17. I am able to break the tension when needed; I can keep the stress within the group at a manageable level.	1	2	3	4
18. I help students see connections between old and new material.	1	2	3	4
19. I take different learning styles into account when I plan a workshop session.	1	2	3	4
20. I treat all students fairly.	1	2	3	4
21. I keep tabs on individual students' progress with the workshop problems.	1	2	3	4
22. I know how to make referrals to other campus resources when necessary.	1	2	3	4
23. I think about the possible impact of race, class, age, and gender issues on students' learning.	1	2	3	4
24. If I had a student with a disability in my workshop, I would know how to plan for support for him/her in my workshop sessions.	1	2	3	4
25. I maintain a positive attitude.	1	2	3	4

Other comments:

Credits

- "Are Our Students Conceptual Thinkers or Algorithmic Problem Solvers?" by Mary B. Nakhleh, which appeared in the January 1993 (Vol. 70) issue of *the Journal of Chemical Education*, on pages 52-55.

- *College Teaching*, 43, 3, 101-105 (Winter 1995) by Gallos, J.V. Gender and Silence: Implications of Women's Ways of Knowing. Reprinted with permission of the Helen Dwight Reid Educational Foundation. Published by Heldref Publications, 1319 Eighteenth St., NW, Washington, DC 20036-1802. Copyright ©1995.

- "Concept Maps in Chemistry Education" by Alberto Regis and Pier Giorgio Albertazzi, which appeared in the November 1996 (Vol. 73) issue of *the Journal of Chemical Education*, on pages 1084-1088.

- Ellis, David B., *Becoming a Master Student*, Eighth Edition. Copyright ©1997 by Houghton Mifflin Company. Used with permission.

- "How Do I Get my Students to Work Together? Getting Cooperative Learning Started," by Marcy Hamby Towns, which appeared in the January 1998 (Vol. 75) issue of the *Journal of Chemical Education*, on pages 67-69.

- National Science Teachers Association. Felder, R. M.: "Reaching the second tier—Learning and teaching styles in college science education. *Journal of College Science Teaching*. March/April, 286-290.

- Perry, W. G. (1970). "The Student's Experience," from *Forms of Intellectual and Ethical Development in the College Years*. Reprinted by permission of Jossey-Bass.

- "Rare Bird," from *Ship Fever and Other Stories* by Andrea Barrett. Copyright 1996 by Andrea Barrett. Reprinted by permission of W.W. Norton & Company, Inc.

- Rosser, S.V. *Re-engineering Female Friendly Science* (New York: Teachers College Press, copyright 1997 by Sue V. Rosser. All rights reserved.), pp. 38-52. Reprinted by permission of the publisher.

- Woods, M. & Blumenkopf, T. A., Stern, V., Swanson, A. B., & Wohlers, H. D. (ed) (1996). *Working Chemists with Disabilities.* Washington, DC: American Chemical Society. Reprinted by permission of the publisher. pp. 56-58.

- "The Workshop Chemistry Project: Peer-Led Team Learning" by David K. Gosser, Jr. and Vicki Roth, which appeared in the February 1998 (Vol. 75) issue of the *Journal of Chemical Education*, on pages 185-187.

Acknowledgments

We would like to express our appreciation to our participating colleges and universities and to the National Science Foundation for their ongoing support of the workshop model.

More information about the model and the support from NSF can be found at the PLTL Workshop website at:

http://www.sci.ccny.cuny.edu/~chemwksp/index.html

The authors would also like to thank Lynn Ashby, Arlene Bristol, Kathy Castellano, Mark Cracolice, Nirmala Fernandes, and Jack Kampmeier for their invaluable assistance in preparing this book.

Index